SECRET
LANGUAGES
OF THE SEA

SECRET
LANGUAGES
OF THE SEA

Robert F. Burgess

DODD, MEAD & COMPANY
New York

Material from *Shark Attack* by David H. Baldridge © 1975 is reprinted by permission of Berkley Publishing Corporation, New York.

Material from *Abyss* by C. P. Idyll © 1964 by Thomas Y. Crowell Co. is reprinted by permission of Harper & Row Publishers, Inc., New York.

Material from *Communication Between Man and Dolphin* by John C. Lilly, M.D. © 1978 by Human Software, Inc. is reprinted by permission of Crown Publishers, Inc., New York.

Material from *Secrets Beneath the Sea* by Robert F. Marx © 1958 is reprinted by permission of Nordon Publications, Inc., New York.

Material from *Scott's Last Expedition* by Robert Falcon Scott is reprinted by permission of John Murray (Publishers) Ltd., London.

1 2 3 4 5 6 7 8 9 10

Library of Congress Cataloging in Publication Data

Burgess, Robert Forrest.
  Secret languages of the sea.

  Bibliography: p.
  Includes index.
  1. Marine fauna—Behavior. 2. Animal communication.
I. Title.
QL122.B89   591.59′09162   81-5544
ISBN 0-396-08011-1   AACR2

TO WILLIAM REISS,
WHOSE SUGGESTION LAUNCHED THIS BOOK

Contents

Preface

When I was very young I developed a strong desire to know what it was like to live underwater. The urge to sit on the bottom of a lake breathing like a fish was so compelling that I often dreamed of it. In my dream it was a wondrous world filled with strange sights and sounds. Even the act of breathing water seemed as natural as breathing air.

I soon learned, however, that the difference between the dream and reality was considerable. This realization came shortly after I carried a heavy rock out into a lake and sank with it into the depths. In reality, the water was only a few feet over my head, but it might as well have been an ocean. There I sat in the mud on the bottom like a Buddha with goosebumps, hugging the rock, my cheeks puffed, my eyes bugged, my ears straining—all senses fine-tuned, ready to receive the rapture of the depths.

Nothing happened. My bugged eyes saw nothing more wondrous than muddy water. It was cold, dark, scary, and deathly still. The only noise I heard was inside me. My heart thumped wildly, my ears popped like distant firecrackers, and when I tried humming to lift my spirits, the sudden sound amplified so startlingly inside my head that I dropped the rock and shot for the surface, gasping.

Once I mustered more courage I tried the trip again. Things gradually improved. Later, though I built homemade diving gear to explore their depths more closely, Michigan lakes always remained cold, dark, sinisterly silent places to me. When Jacques Cousteau wrote his epic, *The Silent World,* we Midwestern divers thought we knew precisely what he meant by that title, though we had never been in an ocean.

Then one day I dived on a Florida reef for the first time and found that it was far from silent. All around me marine creatures grunted, croaked, popped, snapped, or drummed. It was a noisy place. Moreover, some of the fish acted quite strange—they splayed their gills, tilted their bodies at unnatural angles, and a few even stood on their heads. Others lighted up like neon signs, running through vivid, often complex color changes as if unsure which suit of lights they wanted to wear. I learned that in the Pacific the gray reef shark lowers its pectoral fins, arches its body stiffly, and swims erratically just before it attacks. Then, on a coral reef, I saw a two-inch-long tropical fish go through the same posturing with fins ruffled and body arched just before the little rascal darted out and nipped my shin!

Eventually I found out that all these sounds, actions, and color changes were being performed for a purpose. The animals were using these means to signal something to one another. They were communicating. The posturing shark was signaling the same message as the posturing tropical fish. It said, "Stay away from me or I will attack!" The fish standing on its head with gills spread was signaling to another fish at a cleaning station that it was ready for servicing. Here were visual symbols most species understood. Underwater, humans too were using symbols that elicited certain strong fish responses. For example, the widely sought gamefish black grouper may fearlessly approach an unarmed diver, but that same grouper would flee at the sight of a diver with a speargun, the man/gun symbol sufficient to panic most large species that have learned its meaning.

Communication on these levels is basic and often one-sided. But as one moves up into the realm of the more intelligent marine mammals, things change. It is in this broad area that we humans are most interested, especially in the smaller toothed whales, the killer whales, and the dolphins. Here may be intelligence that some scientists say equals or even exceeds our own.

Secret Languages of the Sea is an inquiry into the many mysterious and often bizarre methods used by marine creatures lacking vocal cords to transmit information to others. The term "languages" in the title is meant in its broadest sense to include all the known means used by marine life to communicate, be it chemical, electrical, acoustic, visual, tactile, or sensory. Primitive though some may be, we need to comprehend the simplest before we can hope to comprehend the complex. We should all be concerned by the fact that almost three-quarters of our planet consists of a virtually unexplored world that hosts an alien life about which we know almost nothing. Some of these inhabitants share our own mammalian heritage and may be our intellectual equals, yet man still annihilates many of them and has made only meager efforts to study a few.

This book is about that life, its languages, and the handful of scientists who have learned what little we know about it. While some seek only to better understand these benevolent marine creatures, others are using our most advanced technologies to pioneer research that may provide us with our first interspecies communication: the long-awaited human/dolphin linkup enabling both man and marine mammal to freely exchange abstract ideas and information. Whether or not this will be accomplished should be known within the next few years.

In compiling this information I wish to express my appreciation to those individuals whose observations, research efforts, and sharing of relative data with me for this project made the book possible. I especially want to thank Daniel Hartman, Buddy Powell, and Charles Tally for furnishing me with so many valuable anecdotes and so much physiological information about our mutual friends, the manatees of Crystal River; Perry Gilbert, H. David Baldridge, and Eugenie Clark for their insights into the extraordinary sensory world of sharks; David and Melba Caldwell for their published papers detailing their communication research with dolphins; Jim McMahon for his information on the unusual behavior of schooling fish; Sue Scribner for her observations of unusual fish relationships; Nathan R. Howe for his data on an alarm pheromone response in sea anemones; Carl Litzkow and his wife, Linda, for contributing so much usable material while trying to get me hooked up with a computer; Bob Bain and his wife, Jeannette, for furnishing me with reports of their ichthyological ESP experiences; Maris Sidenstecker for her description of orcas in communication in the San Juan Islands; Don Kincaid for his exciting account of night dives to capture the first living specimens of the Caribbean flashlight fish; Marjorie Parker and Walt and LaVerne Kreisler for providing me with their extremely helpful personal clipping services; David McCray for the use of his excellent photographs of fish being serviced at cleaning stations; Dr. John C. Lilly's Human/Dolphin Foundation for providing me with information on the JANUS Project; Marineland of Florida for providing photographs of biocommunication research that has taken place at their facility; the Naval Oceans Systems Center and the Office of Naval Research, Department of the Navy, who furnished me with photographs and progress reports of the Navy's marine mammal program and research in this field.

ROBERT F. BURGESS

SECRET
LANGUAGES
OF THE SEA

1

Things That Go Bump
in the Night

At the beginning of World War II, when German U-boats freely roamed the Atlantic Ocean and were a threat to our eastern seaboard, the U.S. Navy was prepared in a rather unusual way to detect their presence. Underwater listening devices called hydrophones attached to our submarine cables were so arranged that a shore listener could monitor any one or a series of these units and supposedly detect the approach of enemy submarines before they reached our shores.

What these early listeners were unprepared for, however, was the general noisiness of the underwater world. Water simply sliding over uneven bottom contours created a certain kind of ambient noise. Added to this was a lot of strange clicking, snapping, barking, and grunting that totally baffled listeners. But other noises were more identifiable. The devices picked up the general cacophony of surface traffic— pulsing propellers, the rhythmic engine sounds of conventional ships. What these listeners had to do was to try and sort out of all that racket the sounds associated with enemy underwater craft. Not an easy task considering that no one had a very good idea of what all that extraneous underwater racket was that their devices were picking up.

These matters finally came to a critical point in May 1942 when a real barrage of noise was picked up one night by the sonic detection system at Fort Monroe, Virginia, that was monitoring underwater listening posts guarding the entrance to Chesapeake Bay. From all indications, it sounded as if a pack of enemy submarines were stealthily making their way into the bay. And if that were the case, they would surely wreak havoc with the military shipping in that area.

But was the noise caused by an underwater fleet of submarines, or by some other peculiar phenomenon with which our listeners were unfamiliar? A call immediately went out to the underwater acoustic experts of the U.S. Navy Underwater Sound Laboratory at New London, Connecticut. Intently, these experts listened to recordings of the strange sounds.

The sound levels varied in a definite pattern. The most intense sound occurred at about sunset. The experts listened, then shook their heads. The racket sounded like a bunch of chirping crickets and croaking frogs.

The experts quickly ruled out an invasion by enemy submarines. But that was about all they could do. Some of them wondered if the hydrophones might be picking up the sound of riveting from a nearby coastal shipyard. Or was it some kind of surface wave disturbance?

Both suggestions were finally discarded. The only logical answer left was that something out there, some underwater creature or creatures, were creating the disturbance. But how would one go about checking out this theory?

A young Navy sound technician came up with a bright idea. He remembered how his niece had once silenced a yapping kennel of dogs. "SHUT UP!" she shrieked.

Instant silence, Not another bark was heard.

Well, shrugged the experts, anything was worth a try. The Navy set out to see if it could override the underwater noise by making a bigger noise of its own. Maybe it could be silenced with even more racket, something with a commanding note to it.

The effort began on an evening when the mysterious noise level had reached such a pitch that it was really rattling listeners' tympana. At sea, in the suspected area of disturbance over the hydrophone cable, the Navy pitched a dynamite cap overboard. The explosion produced the desired effect: silence. At least for a few minutes. Then the sound effects started again, almost louder than before. Harried Navy men quickly pitched over four more dynamite caps and got the same results. The overriding bangs only momentarily quieted the underwater noisemakers.

To those in the know, this problem was more serious than it

seemed. Not only was this noise keeping us from hearing enemy submarines, but along with all the sophisticated sonar equipment strung across the bottom were a bunch of mines, the kind that detonated when they heard certain kinds of sound. Moreover, Navy torpedoes homed in on certain sound frequencies. So, along with the threat of enemy submarines our Navy did not need its own mines and torpedoes exploded by noisy sea creatures.

Even at sea our submarines were picking up sounds they could not identify. For example, here is an entry from the log of the U.S. Submarine *Permit:* "0810. Sound [man] picked up an unusual noise....Could see nothing through the periscope on that bearing. This noise sounded like hammering on steel in a nonrhythmic fashion..."

The source of the sound was later attributed to some unknown marine life. Such events happened repeatedly and pointed up the need for more knowledge about undersea sounds and their sources.

The Navy wasted no time. Shortly after the Fort Monroe incident, the Office of Naval Research contacted every scientist they could find who might know something about this subject. Thus began a top-secret investigation of marine life sounds around the world.

During this stepped-up program, marine biologists realized for the first time that the so-called silent underwater world was not quite so silent. We learned quickly that a whole watery world of sea creatures was at times sounding off loudly enough to make our underwater listening posts quite inadequate for detecting the sounds of enemy submarines. The investigators learned that, although fish lacked vocal cords, they were able to make sounds in a completely different way than any creatures on land.

In land vertebrates, sound is generated by what scientists call the "vibrating reed principle": air forced past vocal cords start them vibrating to create sound. This in turn is modified by the pharynx and mouth into various tone modulations. In humans, for example, the use of the mouth, lips, and tongue create the kind of sound patterns that comprise speech. Since air is a relatively thin, lightweight medium, speech and noisemaking comes easily for the land vertebrates. But what about fish? Lacking vocal cords, they cannot employ the vibrating reed principle. Moreover, water is a heavy, thick medium totally unlike air. How do they create sound?

After much research, scientists learned some of these fishy secrets. They found that they were quite capable of making sounds. These marine noisemakers were producing beeps, grunts, groans, croaks, chirps, burps, booms, and bumps that had nothing to do with vocal cords or the vibrating reed principle.

Looking deep within the fish for the source itself, searchers found that some species were sounding off by using certain muscles attached to the walls of their balloonlike swim bladders to vibrate those walls until their bladders resonated in the same way as drums. The sounds were loud and effective and could be made in or out of the water, giving rise to their common name: drum. Other piscatorial noisemakers generate sound by gnashing their teeth. A common and popular foodfish throughout the southeastern coast and the Caribbean commonly called "grunts" (*Haemulon flavoineatum*) gnash their teeth together rapidly to make a sound similar to that created when one runs his fingernail rapidly down the teeth of a comb. To demonstrate that this was indeed the source of the sound scientists placed a piece of cloth between the upper and lower teeth of a grunt and effectively silenced it, despite the fact that the fish could still be seen quietly gnashing its teeth on the cloth.

Some fish are loud eaters, but it is not the smacking of the lips you hear. Any skindiver within earshot of a coral reef can hear the resonating sounds of feeding parrot fish. The slightly protruding front teeth of this species are fused into a hard, sharp, curved surface resembling the beak of a parrot. Hence the name. And since parrot fish apparently find succulent coral polyps delicious, they spend considerable time gnawing away at the reef with their parrotlike beaks, spitting out clouds of rocklike coral as they consume its tender inhabitants.

Two Delaware Bay fishermen were surprised one day in the middle of the bay to hear faint staccato sounds similar to those made by a woodpecker on a dead tree. Since there was neither tree nor woodpecker in the vicinity, they decided the sound had to be coming from the water. "We were drifting up the bay and the sound was heard regularly in varying depths of water," they said. "As the water got shallower, we heard the sound in two different pitches, with the lower tone being followed very shortly by one of a slightly higher pitch."

What the fisherman probably heard was the sound produced by the Chesapeake Bay croaker (*Micropogon undulatus*) , one of the noisiest and most numerous noisemakers along the Atlantic coast. This member of the drum family, ranging up to eight to ten inches long, produces a raucous croak by rubbing straplight muscles against its swim bladder. The larger the fish, the less loud its croak because its swim bladder has grown larger and less resonate. The sounds are most common and loudest during the midsummer spawning season. Two or more fish of different sizes probably accounted for the different tones heard. Though these sounds are some of the loudest ever made by fish, to hear them above water through the bottom of a boat requires quiet weather and calm seas. Otherwise, even a light chop could mask them.

When marine biologists began tuning in on these noisemakers, they found that a single croaker could be heard up to twenty-five feet away underwater. They were soon satisfied that the huge croaker population in the area was probably the cause of the strange underwater noise heard at Fort Monroe in 1942.

Another finny noisemaker, not quite as loud as the croaker, was identified as *Leiostomus xanthurus,* commonly called the "spot" because of a prominent black spot on its tail. This small but flavorful food fish produces a harsh honk audible up to five feet away. The grunts, croaks, or honks of spots, croakers, and other audacious noisemakers often startle the uninitiated coastal angler trying to disengage his hook from their mouth, and sometimes drive him to toss his catch right back where it came from, fearing perhaps that any fish that can croak or honk might next be tempted to bite.

A less "vocal" and more reclusive type of fish that looks about as unlovely as its disposition is the frogfish or toadfish of the *Antennariidae* family. This sluggish bottom dweller with froglike features burps, honks, or croaks in a kind of Morse code. Again, the swim bladder of this fish has been determined to be the source of the sound. To prove it, the investigators removed the bladder from a living fish and found that it was instantly silent. To further prove their point, the investigators immersed the unfortunate patient's bladder in a saltwater bath and stimulated it with electricity. As a result the bladder responded with the toadfish's characteristic sounds.

Certain fish, such as sturgeons, wear their skeletons on the outside of their bodies and are called exoskeletal. They seem to be able to get around about as quietly as a man in a suit of armor.

Despite the fact that their bodies are encased in cartilaginous plates, sturgeon are quiet until they bump into each other. Then, the clattering of two amorous armor-plated creatures must sound like jousting knights.

Some fish such as the sea robin (*Prionotus carolinus*) produce sound with the same kind of mechanism as the toadfish, but it comes out like the rhythmic cackling of a hen, alternating occasionally with soft clucks. Sea robins have been overheard when no other fish were around, which makes one wonder if they talk to themselves. Experts tell us that they create this sound by contracting border muscles that vibrate the fish's air bladder, thus forcing the gas within it through a tiny portion of the bladder's left lobe. The vibrating partition strengthens the sound waves.

At first, the underwater sound investigators found the ocean full of noisemakers, literally a piscatorial symphony of sound, but not much that was musical. Most of it sounded percussive. The real musicians and operatic vocalists would turn up later among the marine mammals. At the outset, scientists were primarily concerned about the high noise level, not the kind of music being made, so the biggest noisemakers got the priority.

The saltwater catfish (*Galeichthys elis*) produces the typical tympanic sound of a beating tom-tom underwater, but its close relative the gafftopsail catfish (*Barge marinus*) prefers sounding off with a sharp yelp or sob. Trigger fish (*Balistes capriseus*), a leather-hided species so-named for its second dorsal spine that serves as a trigger to unlock its rigidly upthrust first dorsal spine, generates a sound by either clicking the joints of its fin spines or by using its fins like drum sticks to beat upon a taut outer membrane to its air bladder, producing a drum effect. The hogfish (*Lachnolaimus maximus*), another saltwater species, aptly named for its rather protruding snout and porcine face, can produce a harsh rasping sound by gnashing its teeth. This is quite different, however, from the rasping gruntlike noise of the squirrelfish (*Holocentrus ascensionis*), which also grinds its teeth, but uses its adjacent air bladder to amplify the sound.

Other tooth gnashers include the long-snouted filefish (*Monacanthus hispidus*), which the investigators found could produce a loud metallic sound by bringing its front incisors smartly together. The porcupine fish (*Disdun hystrix*) was described as having a voice resembling the creaking of a rusty hinge, which it produces simply by rubbing its toothless jaws together.

Divers in tropical waters near coral reefs or rock piles often hear what sounds like distant popping corn. Even after years of saltwater diving I thought this sound was simply in my ears and had something to do with underwater pressure. But underwater sound sleuths soon traced this disturbance to tiny snapping shrimp of the genera *Crago, Synalphaeus,* and *Alphaeus.* These are the most common of some one hundred species of snapping shrimp found in the coastal waters of the world no deeper than 180 feet. These one-to three-inch-long nonedible crustaceans inhabiting coral reefs, weed beds, rock- and stone-strewn ocean bottoms possess a lobsterlike claw that can close with such violence that it produces a highly audible snap. If you were to place one shrimp in a bucket of water, you would hear a sound similar to corn popping. Scientists say this click or pop generated by a single shrimp lasts only one-thousandth of a second. But if you were in the area of a coral reef where perhaps as many as a thousand of these shrimp combine their clicking efforts into one homogeneous sound, it would register on your ear drums like the sound of a crackling brushfire or a sizzling steak. As you move further away from this noise, the sounds seem to blur together into a soft underwater hiss. These shrimp can also emit ultrasonic sound waves that interfere, like static on a radio, with signals sonar operators are trying to receive.

Larger crustaceans such as crabs add to the underwater racket with even louder claw-snapping noises. The spiny lobster *(Panulirus argus)* and the horseshoe crab *(Limulus polyphemus)* throw additional acoustic static into the sonar picture with their characteristic ultrasonic sound waves. Moreover, the spiny lobster can emit loud clacking or clucking sounds simply by rubbing the base of its antennae against its shell. Divers who catch these elusive creatures by hand are quite familiar with their loud clacks and croaks of displeasure at being captured.

Some fish sound off by opening and closing their gill covers with a resounding thud, clap, or booming noise. The bigger the fish, the bigger the noise. This apparently threatening behavior has been observed both in freshwater black bass and in the larger saltwater species such as groupers and jewfish. Diving under a dock once in a lake, I had a large freshwater bass challenge my right to be there. It spread its fins, flared wide its gill covers, moved aggressively toward me, then slammed the gill covers closed with a muffled clap. There was no doubt in my mind what it meant. Similarly, a large jewfish weighing well over one hundred pounds boomed at me from under a coral ledge near Key West and I knew instinctively that both fish spoke the same language. This time, however, it was no mere clap. The concussion was powerful enough to remind me of a toy I had built as a youngster. It was a simple device demonstrating, I believe, someone's idea of an easy way to knock enemy aircraft out of the sky, a subject kids heard a lot about during World War II. Anyway, my working model was an empty two-pound coffee can with both ends cut out. A piece of innertube rubber was then stretched taut over one opening and tied in place. All you did to demonstrate the effectiveness of the weapon was to point it at a candle flame about six feet away, snap the rubber diaphragm smartly, and an invisible vortex of air turbulence shot over and snuffed out the candle. If you wanted to see the mechanics of the device in action, you had your father blow cigarette smoke into the can. It made a great smoke ring.

After the jewfish incident, I knew exactly how powerful a gust had blown out that candle. I had dived down and found the big fellow lurking under the ledge. As I inched closer for a better look, the fish let me know I was close enough. Its fins flared, its eyes bulged, its gill plates opened up like doors, then everything slammed shut with a resounding boom. The concussion snuffed my enthusiasm for any kind of encounter just as swiftly as that invisible vortex had snuffed the candle in my youth. The message came through just as loud and clear as it had with the black bass.

While it took the Fort Monroe incident to get us to listen more closely to fish and other marine creatures, man was already aware that fish were not dumb animals as far back as the pre-Christian era. Aristotle determined that fish made sounds and he recognized that they were produced quite differently from those of land animals. He did not think that fish possessed a language of composed sounds, but referred to them as "inarticulate squeaks and pipings like the cry of the cuckoo." Others, he observed, were found to growl, rasp, or whine in air after they were captured. And certain large schools of fish were said to have made enough noise to have been heard above the surface of the sea. These were extraordinarily accurate observations to have been made 350 years before Christ!

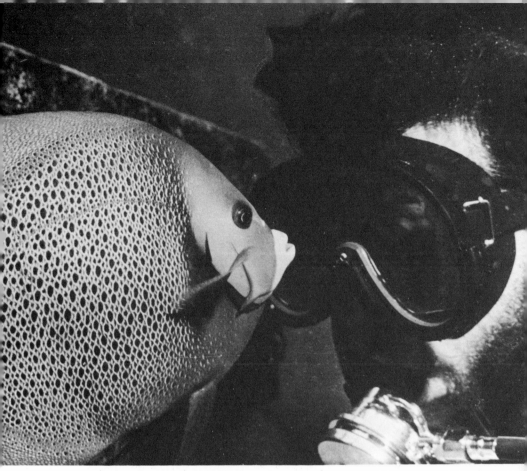

In marine sanctuaries such as Pennekamp Coral Reef Park off Florida's Key Largo, some tropical fish—especially gray angel fish as shown—seem to peer curiously into a diver's mask as if intent on an eye-to-eye confrontation. Actually, the fish are more intrigued by their own reflection in the diver's face-mask than in striking up a relationship.

Despite Aristotle's astute observations and writings on this subject, it was not until the nineteenth century A.D. that man generally acknowledged the fact that marine creatures were actually making noise underwater. And it was only in the early 1940s, during the World War II sound research into the subject, that the importance of underwater sound led us to tune in with a more critical ear on what kind of sounds were being made and by which denizens of the deep. When you consider how many varied noises land animals, including man, can make, you can imagine how perplexing the problem must have been in those years when the scientists first started listening underwater. Their job was to record, analyze, and sort out one sound from another so that sonar men could hear and correctly identify the kind of signal they were listening for. In an effort to describe accurately the chaotic confusion of underwater sounds, the U.S. Navy Electronics Laboratory rounded up some of their sharpest sonar operators, presented them with a multiple-choice list of descriptive words, and let them listen to recordings of the underwater sea life.

Even with trained, experienced sound technicians such as these, they chose a wide variety of words to describe what they thought they heard on the recordings. Understandably, this was no easy task. For example, how would you differentiate between a sound that was either a thump, a bump, or a knock? Examiners soon found that one man's thump was often another's bump. And to still others, it was sometimes a knock! Finally, the only way the examiners could most accurately differentiate between such similar sounds as these was to run each sound through electronic equipment that analyzed its characteristic sound wave and drew a picture of it. Then, a specific descriptive word was assigned that particular sound-wave picture, and forever more it would be known as that. These visible sound patterns, called spectrograms, became the standard method for analyzing and comparing underwater sounds.

During this period of interest in subsea sonics, the necessity for more information on the subject led to the development of better underwater listening and recording equipment. Underwater microphones called hydrophones became more sophisticated and gradually, by the postwar period, things were so vastly improved that researchers were soon working with directional hydrophone clusters that gave them 360 degrees of acoustical scanning. To these were added almost instant analyzers that interpreted what was being picked up, and the data was semiautomatically processed.

In 1954, as more sophisticated equipment for the study became available for use both on land and under the sea, the Office of Naval Research requested the Narragansett Marine Laboratory to establish a reference file of biological underwater sounds that would serve as a current reference library of such data associated with known marine organisms. By 1970 this library had accumulated the characteristic sounds of at least twenty-four marine mammals from the Atlantic and Pacific oceans and the sound patterns of several hundred species of fish ranging from the coastal waters of Canada to the Caribbean Sea.

In the process of compiling all this unique information, investigators soon found out that there were many more underwater creatures than fish responsible for much of the ambient noise they were recording underwater. In widely distributed areas at different times, four phyla of the animal kingdom were largely responsible for all the noise. These included the Arthropoda—the jointed-leg crustaceans such as lobsters, crabs, barnacles, and shrimplike animals; Mollusca—mostly shellfish, but also sea slugs, sea hares, octopuses, and squids; Echinodermata—the prickly surfaced creatures such as starfish, seas urchins, and sea cucumbers; and the Chordata—animals possessing a backbone such as fishes, amphibians, and mammals. Two kinds of western North Atlantic barnacles were found to be responsible for a loud crackling sound similar to that of snapping shrimp: *Balanus eburneus* and his not quite so noisy relative *B. balanoides*. Meanwhile, *B. tintinnabulum*, their southern cousin more common to warmer waters, had an intriguing bell-like name that hardly typified the kind of music this barnacle made by scraping its hard beak across its shell to produce staccato clicking. And across the Atlantic their European cousin *B. perforatus*, basking in the Mediterranean, was playing tunes on its operculer valve by strumming it vigorously to produce what listeners said were "low rhythmic sounds."

No wonder World War II submariners had trouble sorting out their signals from all of that. Later they learned to use this continuous profusion of confused sounds to cover their own activities. And when they found that certain ambient noise occurred only at certain depths over certain kinds of bottom, this information was too good to pass up. Sub crews used such sounds to tell where they were, at what depth, and over what kind of bottom.

The more the experts listened to what was happening in this not-so-silent world, the longer grew their list of classified sea sounds. New England lobsters growl, ghost crabs hiss, and fiddler crabs sometimes make a lot of racket by beating their breasts with their appendages. Black mussels sound off by stretching and breaking their byssus threads anchoring them to rocks or to each other. Scientists discovered that trick in 1958 and ever since they have been calling it "mussel crackle." Certain sea urchins were making a frying sound in New Zealand while their tropical cousin *Diadena setosium* made an even louder noise by snapping its stiff spines.

Most fish and marine mammals apparently sound off loudest during their breeding season. Those living in perpetually muddy water seemed more loquacious than their clear-water brethren. Scientists wondered if low visibility was the reason. Were those swimming in muddy waters like fogbound ships signaling to one another? Or were they sounding off for some other mysterious reason as yet not understood?

To find marine creatures making one kind of sound is marvelous enough. But finding them adding to their "vocabulary" is something else again—especially for the eavesdropping sound classifiers who were having a hard enough time describing single sounds. For example, a certain fish was found to have a basic sound described as a "thump or boom." But when he repeated it, it might then become a "rumble, growl, or drum." Other fish started out with a basic "knock," that evolved into a "croak" and sometimes a "honk." Where one fish might "grunt, cluck, and bark," then repeat the performance precisely, another might "click and snap," then do a "scrape and rasp" for an encore.

Mammals were largely responsible for a lot of clicking. One case of repetitive sound from a mammal started off with a "creak" and ended up with a "carpenter." (This from a captive dolphin mimicking the sound of a carpenter hammering in the area.) Another mammal "moaned, barked, and squawked." Barnacles usually only "clicked," but in some cases their clicks turned to "crackles," as with most mussels.

What is the meaning of all this noise? Do these animals understand what it is they are sounding off about? Or is this simply the unintelligent, purposeless noisemaking activity of marine life? These are questions scientists are asking today. But would it surprise you to learn that as recently as the 1930s at least half of the early investigators believed that fish could not hear? Hard to believe, but true. In an attempt to settle the issue one way or another, researchers were further nonplussed when they removed the inner ears of goldfish and found to their amazement that the fish still responded to sound! How is this possible? Did the fish somehow perceive sound vibrations through their bodies? If this were the case, quite probably then they could not hear in the usual sense of the word, said some researchers.

Other experimenters were quick to disagree. Especially one enterprising individual who attached a bass-viol string to an aquarium, plucked it and declared that 96% of his fish reacted to the sound. Were they, he wondered, hearing the strumming with their inner ears or from vibrations through the water that they were picking up with other receptors? He set about to find the answers by first numbing the skins of his experimental fish and then strumming a note on his bass-viol string. This time, the fish still responded to the strummed string. That seemed to indicate that they were hearing with their inner ears. So he cut the auditory nerves in these fish and deafened them. But were they really deafened? Once again the experimenter strummed his bass-viol string attached to his aquarium. This time he noted that over 75% of the fish failed to respond. The ears had it. Or so it seemed. This was the end of the experiments, but my curiosity would have been aroused by that 25% of the fish that apparently *did* respond. How had they heard without an auditory nerve?

After much proving and disproving of each others' theories, it was finally generally conceded that fish hear not only with their ears, which are often poorly developed organs to begin with, but also with a series of special cells along the midline of their flanks that respond to pressure waves, the kind created by underwater sound. These sensors are easily seen in most fish as a linear demarcation, often the main stripe on each side running from head to tail. Scientists call it the lateral or laterallis line. If you were to magnify these cells, you would find them different from any of the others. The cells are designed to be easily affected by the expansion and compression of water molecules comprising an underwater sound wave. Nerve endings within these cells sense these subtle changes and instantly inform that portion of the animal's brain that interprets what it hears.

Scientists are still unsure how large a part this lateral line plays in directing the animal's behavior, but its mysteries are gradually being revealed. At first, however, it was a major revelation simply to learn that this line of unique sensors was so capable of sensing sound.

What apparently took so long for scientists to agree upon seems to have been a fact accepted by most anglers the world over. While they may not have understood the how of it, they knew full well that fish could hear quite acutely. For centuries, fishermen had used sound to attract fish to their nets. Whether it was the clacking of the coconut shells underwater by natives of the South Pacific or the thudding of bait being chopped up in small boats by mackerel fishermen in the Madeira Islands, both knew that such sounds attracted fish in their areas.

Not long ago some of the most popular items being sold to sport fishermen in this country were lures that supposedly warbled attractive sonic pulses and battery-operated humming devices labeled "fish callers." Legend has it that their origin came about one day when a city slicker, fully equipped with the fanciest fishing gear money could buy, was completely outfished by a youngster using a willow pole, line, and safety-pin hook. Upon seeing the long stringer of fish the boy had caught, the fishless angler asked the lad to tell him his secret. With a grin, the youngster hauled up out of the water a weighted fruit jar containing several furiously buzzing bees. The bees, said the boy, sounded so much like the humming of insects upon which the fish fed that they were attracted to the jar looking for a meal and found his baited hook. Truth or fiction, the idea caught on with tackle manufacturers. And in reality, probably more fishermen than fish were caught with the lures.

After lengthy investigations involving the recording, analysis, and categorizing of fish sounds, together with prolonged studies of piscatorial behavior, most experts today feel certain that these sounds are being made for some definite purpose, and that most of them are not just accidental noises. Some of the reasons why fish make sounds, they say, is to attract the opposite sex, for orientation, as a defense against enemies, and for general communication and intimidation.

These, of course, are sounds that fish make on purpose, not accidentally. It is now commonly believed by the researchers that the lateral line and other skin receptors are so sensitive to low-frequency vibrations that this may be how individual fish in a school keep themselves so well spaced and coordinated in their movements, especially when the school instantly changes directions without a single member out of place.

More recent research with sharks shows the remarkable sensitivity of these animals in responding to artificially created low-frequency sound waves—the kind generated by swimming or thrashing about in the water. Here again these animals are picking up the subtle pressure waves created by the sounds through skin receptors linked to the sharks' highly developed sensory systems used mainly for finding prey.

We now know that some aquatic animals use their sound-producing and sensing organs to orient themselves to their surroundings, especially under conditions where no light exists for visual references or water conditions obscure visibility.

From the very moment we learned that some fish living in the abyssal depths of the oceans never see sunlight, that they spend their entire lives living in perpetual darkness, we asked ourselves: "How?" How did they find their way about? How did they find food? How could they possibly survive? Similarly, how do certain species of cave-dwelling fish that have lived in total darkness so long that they were born blind manage to get around without the sense of sight? If you have ever watched these blind cave fish in an aquarium, you can only marvel at how easily they avoid colliding with each other or the walls of their enclosure, in fact you might find it hard to believe that they are really blind. But if you look closely you will see they have no eyes whatsoever. We suspect that their lateral line and perhaps other sensors are enabling these fish to find their way in the dark. The same holds true for blind cave crawfish and salamanders; all rely on special senses that have nothing to do with eyesight.

Studies of how bats use their built-in radar to find their way through a maze of wires strung across darkened rooms started scientists wondering if this was how some fish navigate in the dark. In 1944, when research by R. Don Griffin first unlocked the mystery of this highly specialized acoustic behavior in bats, he suggested the term "echolocation" to describe this unique ability. It meant that these animals not only possessed an organ for emitting on acoustic signal, but they also possessed an organ capable of receiving and interpreting that signal when it bounced off some object within its range. Humans possess these organs, but we are less adroit at using them in this sense. Still, the principle is familiar to us.

If you stood in a valley before a mountain and shouted, you probably heard an echo of that shout. If the echo came back five seconds after you shouted, the sound waves took two and one-half seconds to go to the mountain and two and one-half seconds to return. Since sound at sea level travels at approximately eleven hundred feet per second, the mountain must be approximately three thousand feet away. As you move closer to the mountain and continue to shout, the echo will return more rapidly each time. If it was dark, or you were blindfolded, you could easily tell not only which direction the mountain lay, but by timing the intervals you would know how close you were to it. This is the principle of echolocation as we believe it is being used in a more sophisticated manner by certain animals.

Marine mammals such as the dolphins and whales are of course experts at echolocation. And we suspect that more marine life than we first thought is using it. Besides many fish, we have found that even some crustaceans such as lobsters are using this ability to either sound out their surroundings or orient themselves to their environment.

A classic example in the use of this sense was cited by investigator R. D. Griffin, doing research aboard a Woods Hole Oceanographic ship about 170 miles north of Puerto Rico. Griffin recorded a series of paired sounds from the depths that seemed to come from some fish moving along beneath the recording vessel. It sounded like a long, drawn-out whistle, followed by a fainter repeat of the same signal. Griffin decided that it was the call of a fish followed by its echo bouncing back from the bottom. Knowing the depth and the interval between the call and its echo, Griffin located the source as being between 1,250 and 3,925 meters below the surface. It seemed unlikely that a single animal would produce the two types of sounds alternately with such regularity. It seemed even less plausible that some second individual could answer the first with the same precise regularity. To Griffin, the most likely explanation was that some unknown animal swimming through the dark depths was orienting itself to the echoes of its own call and therefore was able to maintain its same distance above the bottom. From then on, the unidentified subject of this incident was referred to in scientific literature as the "echofish."

Soon after realizing that marine animals were capable of sounding off for such sophisticated reasons as echolocating, we began to wonder just how capable they really were. Were they perhaps even communicating with each other? Did they have a language of their own? When laymen asked scientists these questions, they were quick to point out that marine animals that make noises do not necessarily do so to communicate. To communicate, they said, required some kind of purposefully arranged sounds that could be emitted and interpreted as "language." Most scientists doubted the ability of these marine creatures to actually communicate in a meaningful manner as we interpret the word *language*.

Others, however, reserved their decision. They had taken the trouble to don diving gear and go underwater to study marine life behavior in its natural environment. What they saw was ample evidence that in fact several forms of communication existed among marine creatures, the same creatures that not many years ago were generally believed to be largely deaf and dumb. What amazed me most of all, I guess, was to realize that fish of many different species were all capable of communicating certain kinds of information in a form of sign language.

2

Body Language

It happened to us one day while diving on a coral reef in the Florida keys. My companion and I were over a sand bottom between walls of coral, hand-feeding a bunch of professional panhandlers. Big fish and little fish swarmed around us to get at the delicacies we were opening for them—black-spined sea urchins. These living pincushions were common to the area, but they were unavailable to the fish until someone came along to crack open their shells and offer them the succulent inner meat. It was like opening hickory nuts for squirrels.

These particular fish had never seen us before, but they were as tame as hand-fed pets in a park. Which meant that they had encountered divers before who had taken the trouble to find and open urchins for them. Still, they acted famished, consuming everything but shell and fingertips. When the fish finished their meal, they did a strange thing. Several of the larger species darted over to the coral wall, seemed to chase their tails a couple of times, darted back to us, then repeated the rush back to the reef.

As I watched them do this two or three times, it reminded me of a fine rabbit hound I used to hunt with as a youngster. Whenever the dog thought he had run a rabbit into a hole, he excitedly came and got me. His prancing told me at once that he wanted me to follow him to the hole so that we could dig out that rabbit.

This was exactly the kind of message I was getting from these fish. Either it was my imagination, I thought, or like my rabbit hound they wanted us to follow them. It seemed they were trying to entice us back to the reef.

20

Curious now, I swam over to see what they were doing. Each fish had stationed itself near a small hole or crevice in the coral. Protruding from these holes were the tips of the long black spines of sea urchins. The fish danced around these urchin holes just as excitedly as my rabbit hound would dance around a rabbit hole. I stared at the fish. Head down, pointing at the urchin, they looked first at me, then at the urchin. If they could have barked, I think they would have. The fish knew better than to try and root out a sea urchin, protected as it was by its sphere of needle-sharp spines, and their request was loud and clear. I responded by digging the urchin out of its crevice with my knife and opening it so that they could enjoy another feast.

This satisfied them only momentarily. As soon as one urchin was consumed, back to the reef they darted to "bird-dog" another meal. How had these fish learned to do this? How had they known that they could "speak" to me in this body language of theirs and that I would respond by doing exactly what they wanted?

We were diving a popular reef where about two hundred scuba divers dive daily. The fish were merely teaching us a trick they had long ago mastered. No telling how many divers they had taught to perform in the same way, and they appeared to be polished performers.

Anyone who has ever spent much time underwater observing fish in their natural habitat, or even watching them in an aquarium, realizes that considerable communication occurs between species, yet no sound has been made. Much of this falls into the category of what we could call sign language or body language. It never ceases to amaze me how incredibly sensitive fish are to the subtlest kinds of gestures.

If they are to be successful underwater hunters, experienced spearfishermen learn quickly never to swim toward or look directly at the quarry they are after. Showing the slightest interest is enough to trigger the flight response in the more popularly hunted species. A spearfisherman learns to approach his target obliquely, watching it out of the corner of his eye, but directing his attention elsewhere, until within range. Then he turns abruptly and fires. With luck he will get what he is after.

Even then, success is not guaranteed. The hunted species are uncannily attuned to the slightest sign of aggression by a

Free or in captivity, fish respond for food rewards. These are sampling a morsel provided by a diver. Others seem almost able to read a diver's mind in making themselves available when food is about to be presented. While some species seem to have developed a sixth sense in anticipating this, it is more likely that they read certain signs—divers finning or fanning the bottom, for example—that tell them that food in the form of burrowing organisms is being exposed.

spearfisherman. Again, this is a learned reflex. Fish have learned to be cautious of man and his weapons from previous experience. The softest click of a speargun being cocked underwater is enough to scramble every grouper within hearing.

The mere combination of man and speargun is sufficient to panic a whole reef full of large fish. Small fish are unaffected by it because they have never needed to fear this combination. Similarly, the large fish that would normally panic at the sight of an armed diver are much less fearful of a diver without a spear.

Fish in the wild are constantly monitoring their environment with every sensor they can muster. What they fail to see their other senses are quick to perceive, and they are experts at being able to pick up the subtlest kind of stimuli that might concern them. So proficient are they at this business that some divers seriously wonder if some kind of extrasensory perception that we, as yet, know nothing about might be involved.

Bob Bain and his wife, Jeannette, dive certain coastal waters where the local fish population never pays any attention to them until the divers enter the shallows carrying a broken shell and start digging in the sandy bottom for seashells. Bain swears that the fish swarm around them from the time they are in the water until they start digging, then they home in on all the morsels of food uncovered by the digging.

Bain asks, "How do these fish know what we are about to do and that our efforts will produce a meal for them? Do they read our minds or pick up some sort of vibes from us? We just don't know and we've discussed this strange thing many times. We have even made control dives, cruising without shelling. Always, it's the same—no shelling, no fish!"

Freshwater artifact divers have had similar experiences. As long as they are engaged in nothing more than looking, fish keep their distance. But the instant a diver starts hand-fanning the bottom for pottery sherds or fossils, the fish are immediately in the hole with him! This reminds me of one more special technique used by successful spearfishermen who find their quarries staying too far out of range. The diver simply drops to the bottom and starts throwing handfuls of sand up into the water, paying no attention whatsoever to the fish he is after. Gradually, the fish's curiosity will get the better of it and it will move in closer to the diver to see what he is doing. And that is often the quarry's undoing. The flying sand is the sign of some creature bottom-feeding, uncovering food, so rather than the fear response to this visual symbol, we have the direct opposite. This man has knowingly used an incorrect symbol to communicate with the fish and elicit a desired response. That is communication, primitive though it may be.

The Bains shared an experience with sea urchins that was of particular interest to me because of the odd fish behavior we had observed with them. The two events were separated by some seven hundred miles. We were in the keys while the Bains were off the coast of northwest Florida in the Gulf of Mexico. Bob and Jeanette had joined a group of other scuba divers to dive on the *Joe Simpson* barge, a wreck offshore from Panama City. Since the wreck is a popular fishing area, it is often festooned with saltwater sinkers and lures. As the group approached the wreck, said Bob Bains, "three of our divers pulled out their knives and proceeded to cut some of these lures and lead sinkers free to drop into their pockets. Now no fish showed up to observe these actions. They all went about their business. However, Jeanette became interested in the sea urchins which were attached to the *Joe Simpson*. She went to touch one of them, but found it too tightly attached. Our dive group leader saw what she was doing and pulled out his knife. Immediately, a whole bunch of fish zoomed in on the scene. He reached over with his knife and cut the sea urchin in pieces, at which point the fish gathered around for the feast. Now why did these fish not appear when knives were drawn to cut the sinkers loose? Yet when a knife was drawn to cut open the sea urchin, they gathered quickly before Jeannette had any idea as to what our group leader was about to do. They *knew* before we even had the slightest idea as to what he was up to. How?"

These small fish photographed with closeup lens are performing a strange learned routine. Having been fed a sea urchin by the author, they swam to the reef and are actively "pointing" to another partially concealed urchin (the black spines at bottom) which they want opened for them to eat. This is a learned reflex, but it is interesting that they have communicated their wishes to a human through body language and he has learned to respond accordingly with the requested action.

Bain described another diver's strange experience that had chilling overtones. He said, "A well-known diving instructor with many years experience in the business told us the following story while we were having dinner after a day of diving in the Bahamas: 'I was spearfishing here in the Bahamas with a buddy in about thirty feet of water and had just speared a large fish which I estimated to weigh between eight and ten pounds. In the ensuing struggle, I lost the fish, which drifted to the bottom. I had broken my speargun during the battle and was so engrossed with the problem that I had failed to notice a large hammerhead shark which had come into the arena. I looked up to find this hammerhead looking at me very intently from about two feet away. Suddenly, I knew that this shark wanted me to give him my fish. It was just like a message had been implanted in my brain, just like someone had said to me, "Get that fish and give it to me."

" 'Well, I was not about to do that. Instead, I placed the butt of my speargun against the shark's head and pushed very gently. The shark backed up a bit, then swam slowly around in a circle and came right back face to face with me. Again, I got the message, "Get me the fish." Again I placed the butt of my speargun against his head and gently pushed. This time the shark pushed back, gently but very positively.

" 'I figured I had had enough of this peculiar encounter so I dropped to the bottom, picked up the fish, and held it out. The shark glided down, took the fish from my hand, circled around and swam off into the distance. I headed for the boat, and sat quietly for a while, pondering what had transpired.' "

If fish are able to communicate by sign language, which in their case is synonomous with body language, what kind of signals are they using and what messages do they convey? As with most land animals using body language, the action usually triggers immediate response in the recipient of such a message. Most of this communication falls into two categories—threat or seduction. Sometimes the human observer of these piscatorial body messages may have difficulty determining whether the fish wants to fight or mate. The colorful fin-splayed, quivering dance of the male guppy trying to attract a female of the species is strikingly similar to the stiff-finned, arched, and swaying posturing of the gray reef shark (*Carcharhinus menisorrah*) just before it attacks.

With its back arched, its pectoral fins lowered and its swimming erratic, the agonistic display of the gray reef shark clearly communicates with body language the warning: "Stay away from me or I will attack!"

Such gestures are basic in this unspoken language and, as with audible sounds, usually bring an immediate response. The audible signal, however, has a more far-reaching effect. The booming of a jewfish commands instant attention. The entire reef population, sensing the sonic signal, pauses momentarily to see what in the world is going on. It has the same effect on this population as did the Navy's explosive caps fired underwater. Everything is momentarily stunned and figuratively looks around to see what is going to happen next.

Similarly, the unspoken signal of body language can command the instant attention of all animals that see it. And the larger the creature making the signal, the greater this response. The flaring of gill covers and fins, the showing of body colors, stiffened or quivering body attitudes—all are used to warn off an intruder. Marine animals recognize these as threatening gestures.

Many fish are territorial and select a certain portion of the reef as their own. Any intruder in this area is met immediately by a flaring of fins and quivering, threatening body movements.

Many times on the Florida reefs I have been attacked by territorial fish not much more than two or three inches long. It usually happens when I am preoccupied in taking a photograph and my swim fin accidentally strays into some little fish's territory. The sharp nip on the shin is an instant reminder that I have trespassed. When I turn around and stoop down to see the assailant, he is often hiding behind a sprig of coral. But as I move closer, he rushes out indignantly with all fins and scales ruffled to make him look as big and mean as possible. Nervously, he darts back and forth, making threatening feints at me, his actions speaking louder than words.

I am always awed by the raw courage of these little fellows so ready to defend their property. Lest I offend them any more than necessary, I always quickly remove myself, and we all go about our business.

Until recently, it was not widely known that some species of sharks are territorial. At least one attack that occurred in shallow waters on Florida's southeast coast was believed to have happened because a boy and his father wandered into what, at least for a moment, was a shark's private territory.

In this instance, the father was towing his son on a rubber float in knee-deep water when two sharks swept in and severely mauled the boy. Scientific shark experts who rushed immediately to the scene found tooth fragments in the boy's legs that were identified as having come from a bull shark (*Carcharhinus leucas*). More significantly, the bite marks indicated the victim had been raked by teeth, rather than subjected to an outright bite. From the evidence, the experts concluded that perhaps the man and boy had come upon two sharks mating or engaged in territorial behavior. The attack was thought to have been a warning, an effort to drive off the human intruders the same way an aroused dog might attack with snapping jaws intended more to frighten than to bite.

One wonders if free-roaming predators establish territories wherever and whenever the mood strikes them. I've known divers to encounter certain specific sharks week after week in the same particular area. In this case it was a rock pile in a normally expansive plain of sandy bottom, and because this was the only protective feature in this otherwise bleak landscape, fish (i.e., food) abounded. This might explain why the shark—a bull—remained there.

Whether it was his private territory or not, divers never had any trouble with him.

Barracudas are great territorial predators, especially the old and venerable that sometimes grow up to six feet long and always appear to have nasty dispositions whether they do or not. Their chosen territories are often prominent underwater structures such as the pylons of an offshore tower, a massive pinnacle of coral, or a shipwreck. In the underwater world, man is one of the largest and most threatening animals in the marine community. Even big barracudas move aside when man appears. At least every one I had ever encountered in many years of diving always did, with one exception. Friends and I had been drifting with the current and free-diving for spiny lobsters near the Marquesas Keys, when I saw the grand-daddy of all barracudas eyeballing me on the very edge of my visibility. He paced me as we drifted, and each time after I made a trip to the bottom I found myself reorienting to his position as soon as I surfaced for a gulp of air. On one such dive, I found he had moved in considerably closer and was watching me with his big baleful eye while he exercised his jaws and flashed his ivories.

Ordinarily, if a diver moves toward a barracuda, it usually manages to maintain the same distance between them or simply swims away. In this instance I moved forward and made a "begone" gesture in his direction.

Instead of responding as any normal barracuda would, this one moved up to within arm's length of me and acted a bit more nervous than I was. At that moment, I had the distinct feeling that this big fellow was trying to tell me something and that something was that I was trespassing.

I promptly backed off then and passed the word to the other divers that there was a 'cuda in the area that was acting less than friendly. One of the other divers took one look and decided that, if we were to continue peacefully, the barracuda had to go. He climbed aboard our drifting boat for his speargun, but when he returned the barracuda had already left. As we prepared to dive again, I glanced down at the bottom and saw that we were no longer over a terrain of sea fans and coral heads. Instead, bottom was much farther away and it was a sterile, white, featureless expanse of nothing but sand.

Rethinking what had happened, I realized that the barracuda had not left. We had. We had drifted completely out of his territory, which probably explains the end of his hostility. He was simply anxious for us to leave.

Normally, large predatory fish of this kind are far less fussy about guarding their territory than are the scrappy Lilliputian fish. Other fish know this too. We often see barracudas drifting almost side by side with the same fish that comprise their daily menu. Somehow the foodfish know that their predacious friends are not in a feeding mood. But let a barracuda's actions suddenly quicken, let it snap its jaws a few times, and there won't be another fish in sight.

A large shark commands instant attention from both man and fish. But again, schools of smaller fish will often feed unconcernedly in the same area as a cruising shark, somehow knowing that, for the moment anyway, the Lord of the Deep's intentions are benign.

How do they know this? What kind of message has passed between the shark and the fish to convince them sufficiently that they are safe? Is it a visual cue? A scent perhaps? Or is it something else—a sound—the sound of the shark's swimming that speaks loud and clear? Just recently, scientists studying shark behavior in the southwestern Pacific found native divers in the Gilbert Islands who could recognize at least three different species of sharks and whether they were "voyaging," "cruising," or "hunting" *from their characteristic swimming sounds!*

Are these the auditory cues instinctively picked up by the small fish and used to their advantage? I would not be surprised.

The gray reef shark, a species that flourishes in great numbers in the South Pacific, is perhaps the most demonstrative of all sharks, using body language to communicate its warning. In 1973, investigators R. H. Johnson and Donald R. Nelson filmed and documented the first systematic study ever made of this strange behavior in these sharks. As the divers approached, the sharks adapted their agonistic display—their backs arched, pectoral fins dropped low, the body held in a stiff "S" shape, and with jaws agape the sharks swam erratically, swinging the forward portions of their bodies from side to side in an exaggerated manner. It was a warning posture, one that the divers could make the sharks perform with greater or less vigor depending upon how close they came.

Continuing this investigation under a research grant from the Office of Naval Research, Johnson, Nelson, and others, diving amid the gray reef shark population off Eniwetok in the Marshall Islands in the mid-Pacific, used a specially designed, bite-proof "Shark Observation Vehicle" containing two scuba divers to see how far they could push the sharks before the display turned from threat to actual attack.

"Using this vehicle, two attacks by gray sharks were experimentally elicited by approaching individuals in agonistic display and partially cornering them," said Nelson. "These attacks were sudden, high-speed strikes prefaced by a display of maximum intensity. . . . Each attack occurred when the vehicle had closed to about two meters, at which point the shark was somewhat (not highly) cornered, a situation which it made no attempt to avoid." Just before the attack, each shark rolled on its side nearly ninety degrees, possibly to orient itself to the aggressive vehicle, then turned and savagely made two rapid bites at the motor/propeller of the vehicle. Nelson noted that in one strike the shark took only 0.33 seconds to approach and the two separate bites occurred in 1.1 seconds, after which the shark swam off again maintaining its display.

Both attacks were filmed for later study and in 1978, as the investigation continued, eight more attacks were elicited from the sharks; these results were being analyzed at the time of the research report.

While this species of shark seems to be inclined toward this combative display, I have watched young lemon sharks under three feet long swim circles around a bottom bait and work themselves into a similar fury of head-swinging, fast, erratic swimming as they seem to crank up their courage moments before darting in to engulf the bait. Other species, such as tiger sharks and great whites, apparently need no such advance self-stimulation before initiating an attack.

In many fish that signal a threat by raising all their fins, the appeasement gesture is just the opposite and occurs when a fish lowers and flattens its fins. Interestingly, since females are often afraid of males, male fish have been seen to modify and tone down their threatening, hostile postures apparently to bring about a more reassuring response in the female.

In the underwater world, color is a language unto itself. I have always wondered how effective such color displays are among the fish in the deeper regions of the ocean where water depths simply filter out the colors of the spectrum. Moreover, experts have long argued over the issue of whether or not fish can see colors as humans do. Or, are they merely able to detect contrasting shades of gray? We now believe that some species of sharks can detect color as we see it because of the color-detecting cones found in their eyes. Since these cones are not always present in others, the ability as we know it may not be the same. But whether marine creatures see color as do you or I is unimportant, providing we accept the fact that they do indeed see and often respond to these different wavelengths of light we call color.

The mere showing of color as a threatening signal among fish was never more significant to me as it was the day a friend and I found a large school of thirty- to sixty-pound cobia *(Rachycentron canadum)* milling around the pilings of a long wood pier on the Gulf Coast in northwest Florida. Not only are cobia extremely fine fighting fish, but they are delicious to eat.

These, however, were most uncooperative. We tried every bait and lure we could think of, but they failed to respond. Finally I ended up just watching them and wondering what they were doing under the pier.

Then another fish attracted my attention, a different kind swimming near the surface parallel to the pier. Since it looked like an elliptically shaped flatfish about two feet long, swimming unhurriedly with a kind of riffle effect of its perimeter fins, I thought it was a Gulf flounder of the family *Bothidae*.

As the fish drew opposite me, three cobia from under the pier suddenly torpedoed up through the clear water toward the inviting target, their mouths opening as they apparently intended sampling this seemingly helpless flounder.

Their abrupt action distracted me from the flatfish for an instant, but in that brief moment I saw the cobia suddenly bolt off in opposite directions as if stunned. Then I glimpsed the "flounder" and saw the reason for their panic. No longer was I looking at a flounder-shaped flatfish but what now appeared to be an expanded version of a bat lying belly down, wings outstretched on the water!

I was almost as shocked as the cobia. Even as I watched, this thing, whatever it was, slowly folded itself back up into the same neat elliptical shape it had, and continued swimming placidly out to sea.

"See if you can catch that thing," yelled my buddy from the end of the pier, where he had seen exactly what I had seen.

Grabbing my surf rod I arched it into a cast. The heavy lead jig splashed down a foot from the mystery fish. It never flinched, just kept swimming on out to sea and that was the last we ever saw of it.

Nowhere, in any of the literature, have I ever been able to identify what we saw. My guess is that it was some kind of ray. Later, comparing notes with my companion, we both agreed that we had seen the thing change shape. But from his vantage point at the end of the pier, my friend had seen something I had missed. He said that when the creature had unfolded, he glimpsed its underside and the real reason, I think, why the cobia suddenly panicked: "Its belly, the side it had flashed at the fish, was fire-engine red!"

Discussing the phenomenon of territoriality with an ornithologist friend of mine one day, he said the behavior was just as common for land animals, especially birds. If you were to sit down in a territory dominated by a nesting male and female mockingbird, you would notice that the birds followed very definite flight patterns around their territory. For example, while the female is on the nest, the male will fly back and forth bringing food to the young, and in doing this he will always take certain similar routes. This is done so that he will not infringe on the territories of other birds. Students of fish behavior have observed almost identical reactions in the territorial fish inhabiting a reef. Such fish roam freely around their own territories, but are reluctant to enter an adjacent territory belonging to another fish.

To leave their territory without crossing another, mockingbirds fly high in the sky, cross over to some pond for water perhaps, then return to the nest in the same manner. Similarly, said my bird-watching friend, fish leave their territories and avoid others by swimming up high in the water mass and crossing the disputed area in this manner.

In a congested reef community, we believe that communications between different species of fish are relatively crude and basic. The only message that seems necessary to be passed from one to another is a warning to keep out of one another's way. These signals are seen constantly between one kind of fish and another of a different species. Occasionally, when flared fins, color displays, and body posturing fails to bring about the desired eviction of the intruder, a fish will add sound to its performance.

William N. Tavolga demonstrated this with a single toadfish whose home was a small conch shell in the bottom of a five-gallon aquarium. Normally the lone toadfish was a quiet, well-behaved fellow, until one day to test his response to this situation, Tavolga introduced another toadfish into the tank.

The reaction was immediate. The first toadfish responded with a series of fierce open-mouthed threatening movements accompanied by sharp grunts that diminished into drum-roll growls. It was a typical territorial defense display.

Tropical fish stores are so aware of this response in certain species that they grade the fish according to how well they are able to get along with their own kind or other members of the aquarium community. Some fish are so fiercely territorial that only one of a kind can be kept in an aquarium without inviting trouble. The Siamese fighting fish is an example.

Scientists have found that color displays and fin and body attitudes play a large part in the courtship and mating of some species. In his study of damselfish, an extremely territorial species inhabiting tropical reefs, a researcher noted that, as a female approached a nest that had been prepared by the male, the male fish would change color and swim rapidly up and down. As it darted down, it emitted staccato chirplike sounds. The researcher determined that this behavior was intended to encourage mating, for when he played a recording of the same sounds to other groups of damselfish, it led to courtship activities. Moreover, he found that each species of fish recognizes its own sound.

When scientists found that certain fish behavior involved the use of colors to communicate, they looked more closely at this phenomenon and especially the mechanism in fish producing these colors. They were particularly interested in how fish could actually manipulate and change their own colors for definite purposes. As a result, they found that a layer of color-bearing cells lies just beneath a fish's transparent scales. These cells usually contain orange, yellow, or red pigments. Other cells contain black pigment comprised largely of the body's waste products. These cells combine with a third type, which creates a yellowish or greenish tinge. And finally, a crystalline reflective tissue adds hues of white, silver, or irridescence. The primary function of color in fish is probably protective. Fish that are dark on top and white or silver on the bottom are virtually invisible when viewed from above or below. Other color patterns on the flanks serve to camouflage the fish against a variety of backgrounds. The flounder, for example, is a master of disguise. This flatfish can alter its color scheme to match pebbles, sand, or dark mud. Many species of fish can rapidly and efficiently change colors at will so that they look entirely different than before. Scientists have found that certain fish have as many as a dozen different color patterns in their repertoire. Some can change colors in seconds. Some species normally colored with black stripes on white have been seen pulling a quick change to white stripes on black within one to two seconds. Others can switch spots and stripes on or off independently. Surely there is a purpose in these color changes, many of which occur during the breeding and courtship activities of the various species.

Certain tropical fish have been observed increasing their color intensity in preparation for combat, while those backing down or signaling a retreat dim their colors. Thus, a fish can literally turn pale with fright.

George Barlow, a professor of zoology at the University of California at Berkeley, described a small maroon-colored fish from India (the *Badis badis*) that literally fights with color. Two fish of this species square off at each other about three body lengths apart, then "fire" color vollies at each other. Staring eyeball to eyeball, they switch on their brightest, fiercest fighting colors until, moments later, one or the other of the fish, without striking a blow, concedes defeat, tones down its colors until it is pale, and swims away, leaving the other fish the victor. What a good, nonviolent way to settle disputes.

Other observers have seen two fish of the same species confront one another in fighting attitudes and fight it out by swift, beating tail movements that direct shock waves to the opponent's lateral lines. Neither fish makes physical contact and the one that loses the battle simply turns tail and leaves. Wouldn't it be great if mankind could resolve all of its battles in this manner? Could it be that the lowly fish have learned the value of mock warfare long before man?

In recent years we have become aware that humans use body language whether they are aware of it or not. Certain positions of the hands, arms, legs and certain facial expressions—a glance, a look, an attitude—are significant signals reflecting our moods and inner feelings far more eloquently than words. When one can read this body language in others, then at least half of the communication link has been forged.

With the exception of people schooled in the use of sign language, some are just naturally capable of talking with their hands. We have all seen them. Scuba divers have developed this silent communication to a surprising degree, especially if they are familiar with each other's body language.

I had never thought much about this until a fellow diver mentioned it one day. Photographing the interiors of underwater caves with little light and handfuls of photographic equipment requires using this kind of sign language under the most difficult conditions. For two years a group of us did this. I had to be able to give underwater instructions to my companions who were either holding lights, acting as models, or doing both. So we learned to use a variety of hand signals, gestures, and mind-reading to get the job done. After one such session, in which a standby diver watched us shoot a picture series, he said later, "Boy, I never saw two guys communicate so much information to each other without either of you saying a word!" And he was absolutely right. Our familiarity with each other's working methods enabled us to carry on a limited but satisfactory exchange of information simply by our gesticulations.

In combination with color changes during hostilities, or courtship, or while tending their young, certain species of fish use seemingly universally understood movements the same way that mimes use exaggerated movements commonly understood by all of us. Thus, in a threatening manner, fish mimic the kinds of things they might like to do to an enemy, but do not actually do. On a reef the Lilliputian territorial defenders are perfect examples of this behavior. To test it, I have slowly moved my camera and closeup lens into a tiny damselfish's territory in a field of lettuce coral and observed a miniature version of the same kind of miming that occurs with fish many hundreds of times its size. The fins ruffle, gill covers gape, and the feisty little fish tries to make itself look larger by puffing up, snapping its tail from side to side like a lioness, darting forward, then back, charging my lens, never really intending to attack it but letting the lens know that here is a vicious little fellow who *could* surely attack it and cause trouble if it is aggressed upon any further.

Adult fish, caring for their young, often call the school together with similar movements, an exaggerated flicking of the fins as if they are about to swim off, but actually do not. In some species where the adults hover above their brood, the rapid opening and closing of pelvic fins discloses a white belly patch and the effect is like flicking a light on and off.

Whether this is a signal intended to summon the brood or simply to flash a warning for the small fry to take cover, we can only guess. But in either case, the young have learned the proper response to these silent semaphore signals. Similarly, among certain cichlid fishes such as the jewel fish *(Hemichromis bimaculatus)*, a small, brilliantly colored freshwater fish of tropical Africa, popular in home aquariums, the two parents use a common sign language in sharing the responsibilities of looking after their small fry, which they keep herded together in a tight, compact school for safety. During the changing of the guard, when the male relieves the female, the male swims in a straight line into the school and performs a rapid zigzag maneuver. Instantly, the female leaves the school, swimming off swiftly in a straight course, while the young, whose attention has been apparently diverted by the antics of the father, do not scatter or follow her but remain in a cohesive school.

Those who study fish that might be considered colorful characters because they use color so frequently in their relationships with other fish soon find that these fish can display "degrees of feeling." Ordinarily, fish are not believed to possess emotions. However, an experienced fish watcher can easily tell just how emotionally upset a certain fish might be by the degree of color intensity it emits. Some fish capable of displaying several color patterns are also able to shade these patterns or "meanings," so to speak. Perhaps they display aggressive colors and actions, but tone them down and thereby combine signals to send out a more complex message. For example, in a stand-off situation between two fish assuming fighting stances, one may be going through all the stereotyped attitudes of tail-beating, body-quivering and fin-arching while already turning pale, signaling that it is soon to give up and retreat. Still others may assume a mottled color pattern across their back to camouflage them from above while their sides take on the bold color stripes warning off aggressors that might attack their flanks. Such two-way signals are broadcasting color-coded messages with quite different meanings.

With all the body language and silent signaling that goes on in the well-lighted shallow water areas of the underwater world, their effectiveness depends on their being seen. Scientists suspect that fish see quite well, that their eyes function much as do those of land animals. However, fish eyes have evolved so that they are highly adapted to the conditions of the watery environment in which these animals live. As we know, fish and other small animals born in the perpetual darkness of underwater caves have evolved no eyes whatsoever. Instead, they rely on other senses to help them find their way about. Conversely, some species of fish that live in the depths of the ocean where light barely exists have developed eyes that are large, light-gathering organs operating on the same principle as a camera lens designed with an extra-wide iris opening that will enable a photographer to take pictures in extremely low light. Some species of sharks, such as the blue shark (*Prionace glauca*), have enormous eyes compared to many others. This may mean that this shark spends more time than most in low-light conditions; perhaps it is a special adaptation for night-feeding or for frequenting the dim pelagic zones of the oceans.

Oddly enough, some of the abyssal sharks that live in the perpetual darkness of the deep and never swim up into the light have eyes that appear more the size of those of coastal sharks. One might wonder why they are not born blind like the cave fish. But even in the deepest oceans there is bioluminescence, chemical light created by other deep-sea fish which comprise the deep-sea shark's diet.

In the light-filled regions of the sea, a fish's vision is largely limited to closeup use. Great distance is seldom important to fish. It is more important that they are able to distinguish color patterns and body language and perhaps the threatening shadows cast by larger predator fish.

The semispherical popeyed shape of the fish eye provides its owner with a wide-angle view of its surroundings for much the same reason that man has learned to use a semispherical glass dome port over his camera lens: to adapt it optically to the watery world in order to achieve the maximum wide-angle perspective. Such a lens is then said to be "corrected" for this effect.

Thus Mother Nature has designed the fish eye so that it is properly adapted and optically corrected to perform this same wide-angle task. A fish's eyes actually bulge from their sockets on the sides of their heads, rather than being located in front as are ours. This side position enables fish to see either forward or backward, covering planes parallel to the sides of their body. Moreover, each eye is capable of focusing independently on objects. These are critical areas for fish, areas from which they might be attacked, and thus they keep them under constant surveillance.

Fish that have been blinded are incapable of using their color changing capabilities for any kind of intraspecies communication. Fish whose optic nerves have been purposely severed by scientists to determine how large a part vision plays in their behavior were unable to change colors to match their environment for protective camouflage.

Much of the foregoing might seem to indicate that fish are quite snobbish to each other, that the species are largely segregated and that individuals set up territories and strive actively to prevent others from trespassing. Actually, this relationship has created a harmonious effect among the different reef animals. Rather than constantly fight among themselves—and this is especially important in such a crowded and overpopulated community as exists on a tropical reef—the inhabitants have solved their potential problems in a remarkably logical manner.

Ornithologists have found that many species of birds inhabiting a marshland or some other isolated habitat have managed to avoid conflict by essentially using the same methods of territoriality as do the inhabitants of the reef community.

This is not to imply that fish or birds never cross one another's boundaries without some kind of altercation, for they do. But it does seem that among the territorial species of both these creatures, most go out of their way to avoid trouble where they know others have established territories. Indeed, some fish have established what might be called "microterritories" for the sole purpose of rendering a service to other fish that come to these territories for this purpose. On a reef, this phenomenon is called the cleaner/client relationship, and it is always a fascinating thing to see.

My Canadian friend Harry Adamson happened upon a unique cleaning station one day as he swam over French Reef at Penne-kamp's Coral Reef State Park off Key Largo. Harry is a great fish watcher. Often he picks a small valley of sand between two walls of coral and simply lies there on his back as quietly and as unobtrusively as possible to watch the passing parade of life around him.

This time, however, Harry was slowly cruising the reef when he came upon several different species of fish lined up as if waiting for something. Harry knew he was looking at a cleaner station.

"They were like automobiles lined up to get a car wash," he said. "A large barracuda fully capable of consuming any one of the smaller species of fish ahead of him patiently waited his turn behind all the rest. At the head of the line was a three-foot-long black grouper (*Mycteroperca bonaci*) getting cleaned by a tiny bright blue fish called a neon goby (*Elacatinus oceanops*). Completely enthralled, Harry got in line and watched.

The grouper's left gill cover was splayed wide open while the tiny blue goby darted in and out, pecking at parasites and cleaning other material out of the larger fish's gills. Finally, as if by an unspoken signal, the big grouper carefully swung its gill cover closed slowly enough to give the goby a chance to dart out. Then the grouper opened its other gill cover for servicing and the little fellow scurried around to do his job there.

Harry watched two or three fish being cleaned in this manner. "And just about the time I was ready to go," he said, "I happened to look over my shoulder and would you believe it—there was a margate and a big barracuda waiting patiently behind me! They thought I was in line waiting to be cleaned!"

The cleaning stations of a coral reef are apparently neutral areas where friends and foes alike may mingle without fear. This kind of operation is repeated over and over on a reef. It makes no difference how many miles separate the stations, the situation is always the same. Any fish, no matter how hungry it might be, seldom, if ever, takes advantage of its cleaner by making a meal of it. This seems to be one of the accepted rules of the reef, though it is not always

The parrot fish *(Scarus gibbus)* is being cleaned by the cleaner wrasse *(Labroides dimidiatus)*. The little cleaners first advertised their willingness to clean by performing a vigorous aquatic "dance." The customer responded by hovering over the cleaning station in an exaggerated posturing. This scene photographed under the Red Sea is duplicated throughout the tropical seas of the world *with the same set of body signals.* Photo courtesy David McCray.

observed. I know of at least one instance where a small, banded coral shrimp *(Stenopus hispidus)* was cleaning the teeth of a large, green moray eel *(Gymnothorax funebris)* and apparently took longer than the eel could wait. In one gulp the eel consumed its cleaner. Any client so uncouth as to eat its dental hygienist deserves a good case of tooth and gum trouble.

Six species of shrimp and over forty species of fish are known to be actively involved in the piscatorial art of preening and cleaning other members of the reef community. Most, however, are wrasses, gobies, butterfly fish, and damselfish. They may set up a cleaning station anywhere near a prominent bottom feature—perhaps part of a shipwreck, a patch of open sand near a natural or manmade object on the bottom, or beside a coral colony. Brain corals are favorites of the small neon goby *(Elacatinus oceanops),* an electric blue and black striped one- to two-inch-long cleaner. Probably the convoluted maze of grooves characterizing this coral affords a good resting and hiding place for the tiny fish that cling there tenaciously with their pectoral fins. More likely than not, if other fish are nearby and look like potential customers, the little cleaner will be out advertising—dancing, darting, quivering, and displaying to its heart's content beside its beauty shop. It always seems that the more it cavorts, the brighter shines the neon blue stripe along its flanks. With its "sign" lighted, no fish in sight has reason to doubt what kind of services are being offered.

The prospective client has only to pull up in front of the station, relax, and wait while the bustling little neon goby and perhaps several of his helper gobies pick and prune, peck and nip at parasites or dead tissue on the customer's skin or gills. If the customer has somehow been wounded, this area receives special attention. The little fish deftly pluck away tidbits of loose and dead tissue, augmenting this menu with whatever parasitic copepods or isopods might be found for the taking.

Meanwhile, the customer lazily presents all sides of his anatomy for cleaning, opening wide his mouth so that particles may be plucked from between teeth, tilting over on one side to present a broader expanse of his anatomy, and in general cooperating to the fullest extent with the busy cleaner.

Sometimes more than one cleaner is operating at a cleaning station. In this case a large fish may be serviced from a variety of different angles by as many attendants as happen to be available. Certainly this speeds up the job and probably makes for an efficient cleaning business.

Communication between cleaner and client is often so subtle that it involves little more than a gentle touch or a color change. If a cleaner wants to get under a fin a client is holding too close to its body, a gentle nudge from the cleaner and the client will lift the fin. Similarly, persuasive probings by the cleaner will induce a client to open its mouth for servicing. Indeed, it sometimes seems that the client fish actually slips into a kind of comatose rapture as it floats motionless, often at a peculiar angle, turning on its side, standing on its head, or even turning completely upside down—all for the sake of making itself more easily accessible to the diligent cleaner.

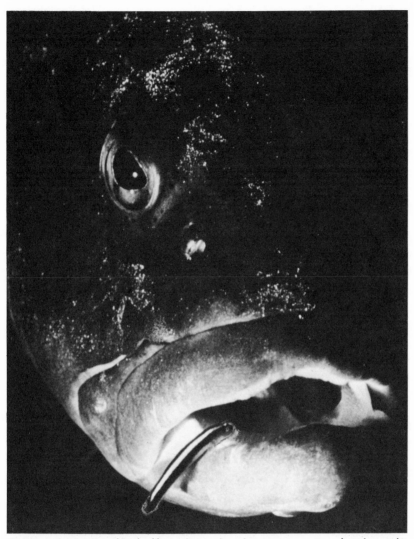

Communicating in a kind of hopping swimming movement, a cleaning goby *(Gobiosoma genie)* services a somber-looking client Nassau grouper *(Epinephelus struatus)* on the Santa Maria Coral Heads of Long Island, Bahamas. Photo courtesy David McCray.

Watching this phenomenon, one gets some idea of the kind of influence the tiniest of creatures hold over some of the ocean's largest denizens. It is always fascinating to see a one- or two-inch goby, or a banded shrimp so diaphanous that almost all you see are the bright color bands on its claws as it manipulates a great barracuda, gingerly moving out along the monster's lips and tapping for entry. As the huge jaws open, the goby flutters into, or the glass shrimp delicately strides into, the tooth-studded mouth and starts to work. The goby picks morsels swiftly in a nervous fashion; the banded shrimp sort of tiptoes through the ivories and delicately, almost fastidiously, plucks tidbits from between the fish's fearsome fangs with all the daintiness of a debutante.

Often the client fish helps the cleaner by changing colors to emphasize particular places where parasites exist or where wounds need cleaning. Certain dark-colored fish sometimes turn pale during the process, while normally light-colored fish may turn dark. Both procedures highlight areas requiring the cleaner's attention.

Underwater photographer Carl Roessler observed an interesting example of this on a reef in Curaçao. Roessler said that, as a school of yellow-tailed goatfish (*Mulloidichthys martinicus*) approached a small coral head that was the cleaning station of a juvenile French angel fish, one goatfish flushed a deep red. Immediately, a tiny cleaner rushed out to this specific fish and serviced it.

Eager to photograph the scene, Roessler carefully swam closer. But the school of goatfish, disturbed by his presence, moved away. Disappointed, Roessler watched them and noticed that the fish that had flushed so red at the station had resumed its normal yellow and white body colors.

By remaining perfectly still, Roessler was rewarded when the school of goatfish returned to the cleaning station and the same goatfish blushed red again. Not even daring to breathe while he took the picture he wanted, Roessler's eventual explosive exhalation again panicked the goatfish into fleeing.

But each time they overcame their fear and gradually returned to the cleaning station while Roessler photographed them. Eight times they left and returned, and eight times the same goatfish turned red for cleaning by the young angel fish. Roessler wondered if the coloration was a request, a kind of truce sign, or whether it was simply rendering the parasites more visible for the cleaner.

The two larger fish shown are purposely striking an abnormal position. With their heads down, tails up, they employ body language to tell the juvenile Spanish hogfish that they are ready to be cleaned. Photo courtesy David McCray.

The spotted trunkfish or boxfish *(Lactophrys bicaudalis)*, a species that wards off aggressors by expelling a potent poison, withholds its lethal toxin while being serviced by the blue-striped lipfish wrasse. On the other hand, a customer eel might occasionally break this trust and inhale its cleaner. But not all cleaners are true-blue practitioners of the cleaning profession. One such phony is the *Aspidontus rhinorhynchus,* which is actually a predator in cleaner's clothing. It looks like a dependable, blue-striped cleaner wrasse *(L. dimidiatus)*; it shakes, shimmies, and advertises as if it were one. But when a parasite-ridden patron pulls up to its cleaner station in all good faith, expecting swift and courteous service, *Aspidontus* rushes out and takes a big bite of it! This little con artist is not a wrasse at all, but a saber-toothed blenny possessing all the ferocity of a pint-size piranha. Characters like the blenny give the cleaning business a bad name.

Despite its occasional back-biters, however, the fish-cleaning business is thriving. During a study of cleaner/client relationships in the Bahamas by the late Conrad Limbaugh, the marine biologist counted over three hundred fish that were serviced by a single cleaner in a six-hour period. Moreover, Limbaugh noted that several of these clients returned more than once during this time to have their fungi infections cleaned.

With typical scientific curiosity, Limbaugh wondered what would happen in this underwater community if all the cleaners went out of business. To find out, he rounded them up from one major station and put them in seclusion while he observed the response.

For a few days, the clients continued to come to the station looking for the cleaners. Then they just drifted off, possibly to other stations; a short while later, Limbaugh reported that the once-crowded area was no longer so. And two weeks after he had collected the cleaners, what few client fish remained seemed seriously in need of servicing. Many showed open and infected wounds, torn fins, and splotches of fuzzy, white fungi growths forming around open wounds.

What is the strange bond that exists between the cleaner and the client that prevents the one from eating the other? No one really knows. But some subtle understanding exists. Fish "requesting" the service of a cleaner have been seen performing all manner of underwater contortions simply to attract the attention of a tiny groomer. Once again, comprehensible signaling. Body language.

In his study of the California kelp fish, Limbaugh reported that the species' popular cleaner, the Señorita wrasse, was never found in the stomach contents of the kelp fish Limbaugh dissected, yet this wrasse is the same size as all the others the kelp fish feeds upon. Is this evidence of understanding among the fish? Do clients recognize and respect the rights of cleaners because of the service they render? So it seems.

Limbaugh also observed that the brown Señorita wrasse of the California kelp beds was such a favorite cleaner that entire schools of clients often swarmed around a single Señorita seeking service. On at least one occasion he saw a Señorita snub its clients as, one after another, they took turns swimming in front of the cleaner and assuming supine attitudes for servicing, while the uninterested Señorita swam nonchalantly off to nibble at something floating by in the water.

Another time, a large kelp fish was seen to present itself in a variety of different positions five times in an effort to solicit the attention of a Señorita that finally deigned to pause long enough to pluck off the client's offending parasite. What observers sometime fail to note in recording this kind of behavior is that the cleaner cleans for one purpose only—to satisfy his appetite. If that appetite is finally taken care of, then the cleaner fish may well lose interest in his waiting clients. Treating them with the same indifference Limbaugh observed, he expects them, one imagines, to take their business elsewhere . . . at least until that particular cleaner once again feels the pangs of hunger. Then, it's business as usual.

Usually we think of the cleaner/client relationship as something that only occurs in the shallow-water fish community. But this is a fallacy. Pelagic or open-ocean species have been seen seeking the same kind of service normally associated with reef fish. Underwater naturalist Hans Hass reported seeing giant manta rays frequently visiting reef cleaner stations in the Red Sea. On one occasion, he saw eight rays swimming around a twenty-five-foot-thick coral formation while awaiting servicing from several hundred cleaners. When its turn finally came, a manta would hover over the formation, splay wide its gills, open it cavernous mouth, and allow the army of cleaners to go to work on it. While some of the little fish swarmed into the mouth, others pecked furiously around the gills. Yet another group roamed up and down the belly, picking and probing for tissue tidbits and tasty copepods.

Hass and others have reported that large pelagic sharks also frequent such stations for servicing. Some cleaners sometimes offer a home-delivery service. For example, the tiny yellow cleaner wrasse, which seems to specialize in sharks, willingly leaves its station and goes into the domain of its client to perform its services, providing the client communicates its desire for such attention by assuming the proper position. Again, body language provides the unspoken request.

Considering the restless perpetual motion of the more active pelagic sharks that must continue swimming to aerate their gills, one suspects that the normal cleaning of these clients is performed by their usual retinue of followers—the ever-present remoras, a species often seen clinging to the flanks or bellies of sharks with the unique suction disks atop their heads. One such remora was seen to run off several cleaner fish so that it could service the client shark itself. To further substantiate this relationship, scientists often find copepods common to sharks in the stomach of these hangers-on. Perhaps the remora/shark relationship is not so one-sided after all. The shark provides the remora with free transportation and the overlooked crumbs of its meals, while the hitch-hiking remora rids the shark of its parasites. Both animals benefit from a common relationship; this is symbiosis at its best.

The entire cleaner/client relationship is to me an outstanding example of communication between species in which the only language used is symbolic and easily understood by both participants. For once, there are no altercations, no territorial disputes. Friend and foe, the hunter and the hunted, all share a common association beneficial to both. Scientists call such relationships mutualism.

Diving with a companion on Molasses Reef off Key Largo one day, I saw my friend become part of this cleaner/client relationship without even the need to resort to body language. He merely placed his hand in a supine position beside a small coral colony. Out pranced a tiny red-banded coral shrimp with absurdly long antennae. Looking for all the world like the most delicately made example of Venetian blown glass, the transparent little fellow climbed aboard

my friend's hand, inspected it carefully, then diligently went about
its business, snipping off flakes of dead skin from around my friend's
fingernails.

As we looked at each other, our eyes smiled because we both had
the same thought: it was a pure and simple case of interspecies
mutualism if we had ever seen it!

3

━━━━━━━━━━━━━━━━━━━━━━━━━━━━━━━━━━━━━━

Silent Messages from Inner Space

Marine animals possess a number of unique ways of communicating that we neither see nor hear. Fish sounds and body language are understandable to us because humans also use these means of communication. We can relate to them. But what about methods that are less obvious to us, perhaps methods that are not obvious at all and are more mysterious than apparent? In humans we might call it a sixth sense, a power of perception seemingly independent of the other five, which enable us to hear, see, smell, touch, and taste. Marine life possesses a similar sixth sense, a unique ability apparently to take invisible messages out of the water itself. With humans we might say such a person is capable of extrasensory perception, of perceiving things beyond our normal senses. This ability is rare in humans, but not in fish. Each and every one of them possesses this gift of extraordinary perception.

Their sixth sense seems to play a part in the way a school of baitfish move as a single unit, its several hundred members able to shift instantly in any direction without bumping into one another. Here, apparently, is instant communication resulting in instant response.

As we know, schooling fish manage this phenomenon by sensing subtle water changes with their lateral line, the row of sensory cells along the sides of their bodies. Never was I so impressed by this precision movement as I was one day while exploring the twisted metal remains of a World War II tanker, the Benwood Wreck, at the edge of the Gulf Stream in Pennekamp Coral Reef State Park off Key Largo.

As I swam up to the Gorgonian-festooned curve of the steel hull, I saw an enormous school of finger-length silver minnows suspended before me like a huge silver cloud. The almost glasslike members of this school were packed together so tightly they almost touched. At my approach, the cloud opened and I moved into the mass.

The curtain of fish closed behind me. I was totally enveloped in the silvery mass of fish all pointing the same way less than two feet from me. As I moved, they moved, the three-inch silver streaks creating an illusion that I was not swimming through water but through a sea of mercury. Top, bottom, sides—all flowed in the same direction, but shifted back and forth depending upon which way I moved. Each shift was performed with such precision by the multitude of silver shapes around me that the effect they created was almost hypnotic.

For a moment I lost all sense of up or down. I had no stable horizons; everything moved up, down, and sideways with the mass ebbing and flowing around me.

Suddenly, in the silvery glare ahead appeared the silhouette of another diver. Instantly, the flowing silver shapes reversed direction and flowed backward toward me, only to halt momentarily, pivot, and move off in another direction. All this was done so precisely, without panic, without a single member of that silver mass one tailbeat out of syncronization with its companions.

Was it really possible that each fish was sensing another's movement with such delicate sensitivity that my eye could not detect anything but absolute precision? Or was some other means of communication involved? And what manner of message was transmitted through the school of fish that diver Jim McMahon encountered in 1977 on the wreck of the *Fenwick Isle* off the coast of Morehead City, North Carolina?

"It was a school of one and one-half-inch-long bluefish in the shape of a perfect ball, not egg-shaped, but an absolutely perfect sphere," said McMahon. "There was nothing rare about this; divers see schools of fish in all manner of sizes and shapes. But as I approached the ball, it split into two smaller perfect spheres, with just one lone fish left in the void between them. The original school was about three feet in diameter with the fish packed tightly together. Watching it split was like seeing a cell divide. The halves were so perfect in size that, had I been able to count them, I believe

Despite the many fish in this dense school of glass minnows swarming around a diver, not one will strike him and not one will collide with another. Scientists believe these feats are accomplished by pressure changes to sensory cells lining both sides of a fish's flanks. This supersensitive sensing system is called the lateral line or lateralis system.

they would have each contained the same number of fish. But why was the single fish remaining alone between the two groups?

"As I backed off, the two balls came back together again around the lone fish. It was all done in so orderly a fashion, as if the whole exercise had been planned. The geometric perfection of the moves would have made Euclid cry. All the fish were aimed at the wreck before they split, after they split, and when they came back together again. Although they were not looking at me, I believe I was the cause of their division. I've often wondered about it. Why the lone fish? Why the two perfect spheres? Was this the best possible shape for gathering information through the lateral line sensors? Could it have been a shape chosen to boost their signal-gathering powers, to amplify toward the center and give all directions equal gathering strength? As for the lone fish, who seemed a focal point, I can't even guess."

Scientists say they still do not know all the secrets about a fish's lateralis system, but essentially, these sensors monitor movement and water flow. Early on in one experiment, to prove or disprove this hypothesis, scientists slipped a kind of rubber hood over the foreparts of certain species of hand-sized laboratory fish. The fish no longer darted away in the presence of a probing hand as they had before their sensory organs were shielded. It was as if they were unaware of the hand's presence in their midst.

Other researchers severed the base of the branch nerve leading from the lateral line to the brain of several research specimens. Again, the fish failed to respond to disturbances in the water and changes in its direction of flow. Still, other factors may be involved in the secret of how vast shoals of fish maintain perfect formation as they swim. Their senses of smell and taste may be factors. These may not be so much intended for savoring food as for savoring subtle, meaningful flavors of the water. Olfactory and gustatory cues may play larger roles in orienting than we suspect. Interestingly, only the lung fish of the order Dipneusti (double-breathers) are known to have taste organs in their mouths. Other fish possess taste buds, but they are located on their heads, bodies and tails. In certain species they are present on long, probing appendages such as fins or "whiskers." One example is the catfish, which is fully capable of savoring water for food far from the actual food source. In the same way, fish probably scent the presence of other members of the school, aiding their other senses in enabling them to follow their companions through the murkiest water, day or night. Moreover, such fish may possess the ability to tune in on their companions and their surroundings by sensing electrical or magnetic forces generated by their movements.

Some of the many sensorlike cells found scattered around the heads and flanks of sharks are now believed to be electrical receptors. Others are believed to be chemical receptors capable of detecting differences in the chemistry of water. Because they are so adept at detecting the most minute traces of blood in great volumes of water sharks have earned themselves the nickname "the swimming nose." How these animals use a variety of specialized sensors to track down prey will be discussed later in more detail.

To these senses, add the ability of certain marine creatures to communicate by light. The most difficult part has been in man's efforts to monitor them. While we know that certain of these animals are able to produce light in a form of flashing signals, we know relatively little about this form of communication. But what little we do know seems to substantiate the belief that these animals do indeed transmit basic information through their selective use of photophores, the light-producing cells located on their bodies.

At a depth of over two miles below the surface of the ocean, where surface light never penetrates, a surprising abundance of marine life exists—creatures that live in perpetual darkness, never seeing the light of day. While perhaps hundreds of these abyssal species never see daylight, their world does have light. Nature has marvelously equipped these animals with their own source of illumination—a cold kind of illumination one might more easily compare to the glow of luminous paint, rather than that of electrical incandescence. Yet it is a light that enables the creatures to see and to be seen.

When naturalist William Beebe descended into the abyss in his bathysphere, he peered through the sphere's quartz ports and saw wondrous things—unearthly creatures bathed in ghostly blue lights, creatures of the deep that assaulted his senses with a display of pyrotechnics exceeding his wildest imaginings. As he was later to describe the experience in his book *Half Mile Down* (Duell, Sloan & Pearce, Inc., 1934), Beebe noted with unrestrained excitement:

I watched one gorgeous light as big as a ten cent piece coming steadily toward me, until, without the slightest warning, it seemed to explode, so that I jerked my head backward from the window. What happened was that the organism had struck against the outer surface of the glass and was stimulated to a hundred brilliant points instead of one. Instead of all these vanishing as does correspondingly excited phosphorescence at the surface, every light persisted strongly as the creature writhed and twisted to the left, still glowing, and vanished without me being able to tell even his phylum.

Moments later, at twenty-one hundred feet, Beebe was enthralled by two strange specimens of barracudalike fish he judged were at least six feet long. They possessed a single row of strong blue lights along their sides. They had large eyes and toothy, undershot jaws, their sharp fangs glowing blue as if from luminous mucous or lights within their jaws. As the long fish passed a few feet from the sphere's window, Beebe saw that each trailed two long tentacles tipped with a pair of luminous spheres of colored light, one red and one blue. Beebe named the unknown species "the Untouchable Bathysphere Fish."

Today, biologists believe that over 75% of the marine species below 2,500 feet of the surface intermittently produce light with photophores. Many of these light-producing cells are located not only around the eyes of the animals, but also along the lower parts of their anatomy, enabling them, it is believed, to see objects better on the ocean floor. Most of this light activity occurs where sea life is most abundant, an area called the twilight zone between 750 and 2,500 feet below the surface. Those marine animals producing this cold light through a chemical process do so primarily to attract prey. Some, such as the angler fish, possess light-producing organs on the end of a slender fishpolelike appendage, which when dangled before their wide-open mouths, entices victims within easy striking distance. Other species with photophores in their mouths lie in wait with their jaws open until some unsuspecting creature, fascinated perhaps by the eerie blue lights surrounding the lethal teeth, is lured into the trap.

Biologists believe that these light-emitting cells are also used for communication. Experts tell us that creatures of the twilight zone flash their light signals, in the same manner as fireflies, for the purpose of mating. Entomologists who have studied fireflies and their luminous love language have become so knowledgeable on the subject that they can look out across a dark field at night and identify a dozen different species of fireflies simply by timing the different flash patterns of their winking taillights!

They have learned, for example, that the male is the one doing all of the flying and much of the flashing, while the female usually remains on the ground ready to respond in kind to a signal that attracts her. Observers have found that the female never flashes her taillight except in an answering response to that of the male.

Timing is everything. The duration of a male's flash is always the same, as is the interval between flashes. The female of one species waits exactly two seconds before returning the signal. If the male is interested, he may circle overhead and exchange coded messages with the female for up to ten times, repeating the same sequence of flashes and intervals before the male finally joins the female on the ground to mate.

In Burma, a certain species of firefly has been observed performing this ritual in unison with others of the same sex: all the males sit in one tree and flash their love lights simultaneously, and after a proper interval, a maidenly glow lights up a nearby tree in which all the females are sitting.

In his book *Abyss* (Thomas Y. Crowell Co., 1964), author C. P. Idyll describes the glowing mating ritual of the fireworm *(Odontosyllis enopla),* common to Atlantic and Bermudan waters, as being perhaps the source of the mysterious lights Christopher Columbus reported seeing in the depths as he approached the Bahamas on his first voyage in 1492. Columbus described these innerspace UFOs as resembling the lights of moving candles. Idyll believes that they may have been fireworms:

> Two or three days after the full moon, the females rise from the burrows in the bottom of the sea to swim in circles at the surface, glowing brightly green. Males in the vicinity that see the circles of light swim toward them, emitting flashes of light themselves as they go. If no male appears, the female turns off her lights and

stops swimming in a circle for awhile, then in a few moments, lights up again and resumes her dance. If the female darkens while the male is approaching, he will stop swimming toward her, wander aimlessly, and only advance again when she luminesces once more. Several males may be attracted to a single female, and the whole group then swims in a fiery circle. Eggs and sperms are emitted simultaneously, and the fertilized eggs float off to create the next generation of worms.

For reasons known only to Mother Nature, some creatures of the deep wear a suit of lights that would shame the gaudiest matador. For example, the deep-sea prawn *Sergestes prehensiles*, which sports over 150 winking, blinking, greenish-yellow light-producing cells over its entire body, is virtually a one-prawn light show by itself. In his book *Bioluminescence* (Academic Press, 1952), a fascinating account of a lifelong study of living light, Dr. E. N. Harvey describes this miniature shrimp-sized fireworks display by reporting that each single light cell flared brightly momentarily, and as the glow diminished, others gleamed anew, but seldom was more than one light glowing at a time, creating a kind of miniature pyrotechnic display with diminutive skyrockets firing into brilliance every one or two seconds. At other times, this remarkably luminous prawn withheld some of its brilliance from certain parts of its body to fire off simultaneously entire banks of photophores, lighting up whole sections of its anatomy. One wonders not only what this prawn does for an encore, but what all these flashing lights signify.

A similar gaudy, cold light performance given by the small Indian Ocean diadem squid, was described by marine biologist Carl Chun over seventy years ago: "One would think the body was adorned with a diadem of brilliant gems. The middle organs of the eye shown with ultramarine blue, the lateral ones with a pearly sheen. Those toward the front of the lower surface of the body gave out a ruby red light, while those behind were snow-white or pearly, except the median one, which was sky blue."

If we do not know why such creatures flash a multiplicity of lights, we are equally mystified as to the reasons some species seem to have departed from the normal yellow, blue-green light range of most of them and, instead, light up with almost every color of the spectrum.

The lantern fish *(Myctophidae)*, a large group of relatively small fish seldom averaging longer than a man's middle finger, but found to range through the darkness of the depths down to thirty-five hundred feet, are adorned with lights. So complex are their light patterns that they resemble small, swimming, neon signs ablaze with a variety of gemlike lights. Pale green seems to be the most popular hue; however, observers say they often see the species glowing in pale yellows tinged with red. Apparently the distribution of these photophores on the lantern fish's body is what quickly differentiates one kind from another. The males seem to have the most and the brightest light-producing cells. In fact, Beebe noted that one lantern fish in his aquarium had a light brilliant enough for him to read by.

On the other hand, the females of the species are less well equipped with light cells. It has been said that this special coding of light-producing cells on the different kinds of lantern fish makes it possible for those studying the species to recognize up to 170 different members of the clan simply by their light codes.

I think this is a prodigious feat. If humans can do it, you can be sure that lantern fish know how to read these identification signals. Quite probably, then, these light emissions are used not only to illuminate the lantern fish's way through his dark world, but also to attract members of the opposite sex for mating purposes.

It is interesting to note the similarity of messages transmitted by two different kinds of fish—those living in the shallower waters in sunlight zones and those whose entire life is spent in the darkness of the depths. Where it is light enough for the inhabitants to see each other without benefit of their self-generated illumination, color becomes the unspoken words, the "language" of the area. We know how color and color patterns are used by both the shallow-water fish and those of the deep for mating purposes. Similarly, other information may be transmitted in which the message is identical—only the method of its transmission varies. For example, a fish living in the shallow-water reef community may flare its fins and display its vivid colors, perhaps even turning on a few one might not normally see, to announce to some potential aggressor that that particular spot is its territory and it dislikes trespassers. In the same manner, one of the species that perhaps has never seen daylight lives so deeply that no color exists as we might normally perceive it, goes through the exact procedure of its shallow-water cousin—flaring its fins and turning on its body lights in a blazing display announcing to the piscatorial world: beware of this garishly lighted apparition.

In both cases, the degree of display may indicate to what degree the animal intends to pursue its threat. Brightness of colors or intensity of light denotes the animal's proclivity for fight or flight. Conversely, fish wishing to make themselves as inconspicuous as possible simply switch off their displays. On the shallow-water reefs at night, fish that are normally vividly colored during the day may be found tucked away between the branches of coral formations performing the equivalent of resting or sleeping with their daytime colors dimmed so drastically that they are pale, colorless caricatures of their former selves. And surely, for the same reason, deep-sea fish simply switch off their lighting displays and blend in to their world of blackness. Whether they are shallow- or deep-water types, they have both turned off and turned in.

In each case, these fish are dimming their colors and their lights for protection, to avoid interesting possible night-marauding predators. There is a time to be seen and a time not to be seen.

Squids and octopuses share the same kind of defense mechanism. When approached by an aggressor, they emit a cloud of black or sepia-colored ink and jet off, leaving the confused attacker contemplating whether or not to proceed. But what good is a squid or octopus's smoke screen in the deep-sea world of perpetual darkness? The ink cloud is of no benefit if the predator cannot see it.

In the same way that nature has provided deep-sea creatures with lights for specialized purposes, it has solved this problem for the deep-sea squids and octopuses by providing them with a smoke-screen that glows luminously in the dark behind them as they jet off with all their other lights dimmed.

The same kind of trick is used by the deep-sea prawns that Beebe described as seeming to explode before his eyes. They, too, left behind a luminous glow of their presence while they escaped into the darkness. Some of the luminous clouds continued to sparkle with pinheads of light long after their originators had vanished. For a while, Beebe was puzzled by the bathysphere's lights illuminating this grayish mist where once there had been a shrimp, until he realized this small cloud was the burned-out material that had been ejected by the long-gone shrimp.

Without a doubt, the greatest light-show experts in the world, the real virtuosos of colored-light manipulation, are the encephlopods, the octopuses, and squid. What kind of messages these "head and feet" animals transmit can only be guessed. Some are obvious, while others are as puzzling to observers as the radio signals received from the Milky Way. And because these deep-sea inhabitants are also aliens of a faraway innerspacial world, we may not soon find out. But from what observers have seen in the shallow-water species, we assume that much the same occurs to their relatives living in the most profound depths of inner space.

Each September, which is springtime in the southern hemisphere, the squid swim into the shallows of New Caledonia to mate and lay their eggs around the reefs. One day, during this period, underwater photographer Douglas Faulkner came upon a lovely female squid hovering over the coral of a reef near Amédée Island. "I cautiously approached to photograph her," said Faulkner. "While I was doing so a male glided up to her and began flashing neon blue colors over his body in waves. He approached still closer and spread his arms wide. With his two top arms he stroked the female's forehead until she spread her arms to receive him. Within a few moments they locked in an embrace. Thus joined they glided gracefully over the reef, making me think of lovers floating over the rooftops in a Chagall painting. Anthropomorphic or not, my impression was of something definitely tender."

After seeing how skillful shallow-water squid can make intricate color pattern changes at will, one might well wonder what prodigious lighting spectacles occur among such family members as the giant squid (*Architeuthis princeps*), one of the rarely seen creatures of the abyss known to attain a length of at least fifty-five feet and to be capable of engaging sperm whales in life-or-death wrestling matches.

Some members of the squid family possess a curious array of multicolored light-producing cells. We know this because many of them have been dredged up from the deepest oceans by scientists using special trawls. Some of these animals have bizarre photophores in parts of their anatomy.

Deep-sea squid expert Gilbert L. Voss, of the Marine Laboratory

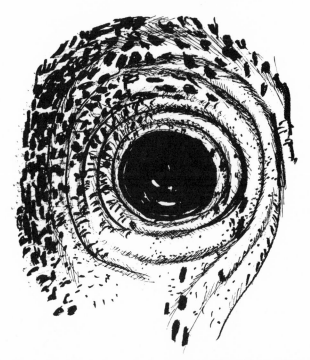

This closeup of a squid's eye shows the pattern of bright red pigmented cells appearing like spattered paint around the eye. Such cells play a role in the squid's very existence. A master of disguise, the shallow water species can switch protective coloration almost instantly. Where color is discernable in the sea, in the visible spectrum ranges, squids use it in their mating ritual in the same way their deep-sea counterparts living in darkness use luminous photophores, colored light-producing cells. Many located in the eye areas are apparently there for the same reason as the pigmented cells of the shallower water species.

at the University of Miami, has described some of the peculiar modifications he has found in these deep-sea squid. One species whose eyes are mounted on stalks extending out from its head has light-producing cells encircling its eyes. Others possess inordinately long tentacles with powerful searchlights on their extremities. Yet another kind has tentacles lined with photophores and hooks. Voss believes that this particular squid probably lures victims within reach by jigging its tentacle lights up and down in a tantalizing manner, the same way a fisherman might "jig" lures. When the victim swims close to investigate the odd sight, the squid skewers it on one of its barbed tentacles.

Yet another kind of squid, whose body is covered with brilliant blue and yellow lights, possesses a left eye larger than its right. Well-defined photophores ring the big eye, but smaller, almost rudimentary light-producing cells surround the smaller eye. Why the difference? Voss speculates that perhaps the larger eye with its better-developed light cells may be used in the darkest regions of the abyss, while its smaller eye and less important photophores serve the creature in the better-lighted upper reaches of its range.

The bioluminescence or chemically produced cold light commonly found in certain fish and squid of the ocean's depths is sparked by the kind of chemical reaction that produces this light in fireflies. But with the fish and the squid, it is truly a "living light," produced in these animals by billions of bacteria living within the photophores. These microorganisms, fueled by the food and oxygen supplied by the animal's blood, respond with a continuous emission of light. In some squid these light-producing cells are covered by a layer of transparent, pigmented skin cells enabling the squid to vary the color and intensity of the light emitted by the photophore.

Certain squid have photophores embedded in the walls of the eyes themselves. We have all seen animals whose eyes seem to glow in the dark because they possess certain optic structures that reflect light so well. But in the squid, the eyes actually do glow with their own internal lights. Other members of the cuttlefish family have these light-producing cells within their ink sacs. And certain kinds of deep-sea shrimp have them in their intestinal organs; the light glows through the animal's translucent body.

Deep-sea fish have the cells spotted in a variety of different places. Some are located within the jaws to create a kind of diffused lighting around the teeth, while others are located along the lower sides of a fish's flanks, the light diffusing through a lenslike translucent scale. One wonders if these are used sometimes by the fish as landing lights on the bottom.

Of all the fish possessing these light-producing cells, one of the brightest luminaries in this group is *Photoblepharon*, commonly called the flashlight fish. These goldfish-sized fish possess large pockets of light-producing bacteria under their eyes so that the photophores shine like headlights. Because of this effect, French divers in some parts of the Indian Ocean nicknamed the fish in their area "le Petit Peugeot."

One wonders how a colony of bacteria could possibly create sufficient light to provide a directional beam, shining forth in front of the fish like headlights. But apparently, elements within these pockets under the fish's eyes—reflective crystals that concentrate the light like the mirror of a telescope—intensify and reflect the light emitted by the bacteria. Equally interesting, these fish can actually control the amount of light emitted by masking the light pouch with an opaque layer of skin cells that lifts to cover it like an inverted eyelid, as does *Photoblepharon palpebratus* of the Red Sea and Indian Ocean. Or, it rotates the eye pocket into a cheek pouch as occurs in *Photoblepharon's* close Indo-Pacific relative *Anomalops karoptron*.

In the New World, two species of flashlight fish—*Kryptophanaron* (Hidden Lantern) *harveyi*, found in the Gulf of California, and *Kryptophanaron alfredi*, of the Caribbean—rotate their light organs downward, but must assist this masking with the help of a lower folding type of eyelid.

In the Indian Ocean, both species of flashlight fish, *Photoblepharon* and *Anomalops*, are common to the waters around the Banda Islands, where the local fishermen have found a unique use for the light-producing organs. They remove them from the fish, hang the luminous pouches on a hook, and use them as a luminous lure to attract unwary fish in the same way that the deep-sea squid jigs its lighted tentacle up and down until a curious prey approaches and is skewered on one of the squid's tentacle hooks.

The ability of the flashlight fish to control the amount of light emitted suggests that this species uses its headlights for more than just finding its way around the ocean's dark depths. Observers of this light have described the brightness as equal to that of a penlight with a weak battery. Scientists say that this fish produces perhaps the most intense light known to be generated by a multicellular luminescent organ. Though we are only recently beginning to learn about this relatively rare deep-water fish, ichthyologists now believe that the flashlight fish not only uses its luminous eye pouches to signal members of its opposite sex, but also for purposes of communication. This fish's luminous eye pouches are also used as a defense mechanism. When confronted by an aggressor, for exam-

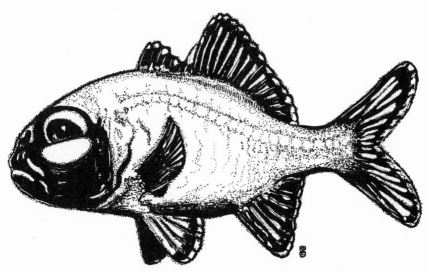

The rare Caribbean flashlight fish *(Kryptophanaron alfredi)* communicate by blinking their luminous eye pouches. Note too the luminosity of the fins. (After a photo by Fred McConnaughey.)

ple, the fish may shut off its lights, swim rapidly up into the face of the enemy, and abruptly turn them on again, a tactic that supposedly frightens off the intruder.

A variation of this theme is called the blink-and-run tactic. When being pursued by a predator, the fish turns on its lights, swims rapidly in one direction, luring an attacker to follow, then suddenly shuts off its lights and darts away in another direction, throwing off the pursuer.

According to ichthyologist Eugenie Clark, who for the last decade has been involved in researching various species of fish in the Red Sea, *Photoblepharon* was first found there by marine naturalist David Fridman in 1964. This species averages about three inches in length and may range from surface depth down to several hundred feet.

Diving among schools of *Photoblepharon* without lights to guide them for fear of frightening off the fish, Clark and her companion, underwater photographer David Doubilet, described it as a sensation akin to being suspended in the total blackness of outer space while surrounded by hundreds of half moons that blinked on and off once every twenty seconds.

"If startled," said Clark, "the 'moons' might wink on and off up to seventy-five times a minute." Moreover, if the school scattered in its usual blink-and-run tactic, observers were treated to the inner-spacial spectacle of half moons shooting off into every direction, blacking out, then appearing elsewhere on a tangent course. Surely pretty heady stuff for these underwater fish watchers. I can't personally think of a more exciting experience than diving blind without lights in the Red Sea surrounded by blinking half moons of luminous light, unless it would be to have accompanied my friend Don Kincaid and his companions on a midnight dive down a vertical reef wall to over two hundred feet in search of the rarely seen Caribbean species of flashlight fish.

For many years, the Caribbean version of this fish, *Kryptophanaron alfredi,* was believed to exist throughout the Caribbean; but it was so rarely seen by the few divers who cared to descend into the depths at night without lights that no one had ever captured a live specimen.

The first of its kind was found floating dead in Kingston Harbor, Jamaica, in 1907. This specimen was collected and preserved, but later lost. Other flashlight fish were supposedly sighted, but it was not until 1976 that a couple of fishermen in Puerto Rico found one of the fish in a trap they had set six hundred feet deep. This one was given to a marine biologist at the University of Puerto Rico, who identified it as the same kind of fish as that found in 1907.

The idea of finding and photographing the first living specimen of the Caribbean flashlight fish originated in Tahiti at the end of 1977. It was there that author Fred McConnaughey, diver-photographer Don Kincaid, and others had gone to photograph sharks. McConnaughey was writing a book on tropical fish, photographing all the specimens that a diver would commonly see while diving in tropical waters. When he mentioned to Kincaid that he was interested in locating living specimens of the Caribbean flashlight fish, Kincaid suggested that they try their luck in Florida and Caribbean waters where he had heard they could be found.

The spot they settled upon to search that January was the dropoff at Grand Cayman Island. This site was selected for two reasons: it was a vertical dropoff, a wall that began at 110 feet, and it went straight down into the abyss 6,000 feet below. If the species of flashlight fish known to frequent the 600-foot depths was in the area, perhaps it would come up this wall at night searching for small copepods, isopods, and other shrimplike crustaceans that rise toward the surface on dark nights. Their diving guide, Captain Paul Humann, skipper of the *Cayman Diver*, enjoyed diving at night without lights. This pursuit paid off for him one particularly dark night when as he was enjoying the bioluminescent view on the wall at 110 feet, the first Caribbean flashlight fish ever seen alive appeared. To Humann, the bioluminescence from the revolving eye pouches of these fish looked incredibly bright in the Stygian darkness. The fish Humann saw, however, were lost when they performed their blink-and-run tactics and disappeared into a cave on the reef wall. Humann felt that, if the group dived long enough in this area, the divers would soon sight another flashlight fish.

Thus began a week of night dives without lights down the vertical reef wall, Humann leading the way followed by the others—McConnaughey armed with strobe light and closeup camera equipment for getting the essential pictures he needed for his book; ichthyologist William F. Smith-Vaniz, of Philadelphia's Academy of Natural Sciences, the man who would identify and capture the species if they found it, armed with catch-net and plastic bags; and Kincaid bringing up the rear with strobe and wide-angle camera equipment for recording the event.

Night after night for almost a week the divers descended down the face of the vertical reef wall in total darkness. Only a trail of bioluminescent blue fire and glowing bubble clusters marked their descent into the depths.

At 110 feet they paused and waited, hardly able to discern each other in the liquid darkness. Far above, they saw the silhouette of their dive boat. A full moon was rising. Its light would make it even more difficult to approach the wary fish they sought, even if they found it.

Failing to see any telltale glowing headlights at 110 feet, they stayed within the shadow of the wall and swam deeper, pausing finally on a ledge at 210 feet below the surface.

Here they were in complete and absolute blackness. All they could see were their glowing instrument panels and wristwatches.

Their bubbles soared upward, gleaming brightly in the eerie blue-green light of disturbed bioluminescence.

At this depth they all felt the effects of nitrogen narcosis. But suddenly, Humann spotted the half-moon glowing eye patches of a flashlight fish. As planned, he snapped on his own underwater light and shone the beam full upon the fish, hoping to stun it into staying still. Instead, the specimen blinked and ran, disappearing into a nearby coral cave.

Unable to remain much longer than ten minutes at this depth, the divers started toward the surface, pausing at certain depths for predetermined lengths of time to decompress, ridding themselves of the nitrogen buildup in their system.

The following day they motored out to the same area and dived down to 225 feet in the daylight to orient themselves to the underwater vertical wall typography and caves so that they could better understand what they were doing in the pitch-black darkness of their nighttime dives.

On subsequent nights, however, they found no fish, even though they sometimes made up to three dives a night to these depths, spending three hours of surface time between each dive in order to rid themselves of the residual nitrogen that could cause the bends.

Finally, with only two nights to go, bad weather forced the group to move its vessel to another location. There, in the dark of the moon, they swam down to the dropoff to enter a cave at a depth of fifty-five feet. Again, without lights, they slowly worked their way through this stony tunnel, their eyes adapted to night vision, their orientation guided only by the bioluminescence surrounding coral outcrops, sea fans, and fish around them.

"The bioluminescence was incredible," said Kincaid. "There were pinwheels and firecrackers going off all around us. Whenever we moved, the disturbance lighted up the water as if we were surrounded by green fire. It was really spectacular. Whenever I touched the bottom, everything lit up. It was like flying over a city at night. You know how you can see the pinpoints of light, all the traffic and street lights? Well it was all there. That's exactly how it looked as we went through this cave."

Finally the divers exited from the cave on the wall at 120 feet and there the four of them sat and waited.

Suddenly in the darkness a pair of half moons winked on and

gleamed brightly. To the four men who had waited so long for this moment, they could not have been more awed had they found themselves in the presence of some luminous-eyed innergalactic creature.

But only after the initial shock of recognition wore off did the four spring into action. Instantly, they surrounded the specimen. Using the flashlight fish's own tactic, Humann transfixed it in the beam of his underwater light. McConnaughey blazed away with his strobe, taking closeup pictures. Smith-Vaniz stood by with poised fish net while Kincaid joined the fray shooting wide-angle photographs. In the ensuing excitement everyone elbowed each other while trying to get what he had come after. Finally, his patience exhausted, Smith-Vaniz swished out with his net, expertly scooping up the prize, and the picture-taking session was over.

Several hours later that night, on another dive, Kincaid separated from the others and chanced upon two flashlight fish beside a gorgonian. His efforts to mesmerize one fish with his dive light, while at the same time trying to take its picture with his camera in one hand and strobe in the other, defies description. Eventually he accomplished the feat by clamping his dive light between his knees while photographing. Then, lacking a catch net, which he surely could not have wielded anyway, he did the next best thing: caught the fish with his hand!

By night's end, the jubilant four found they had captured a total of six lively specimens of *K. alfredi*. But since they had come unprepared for keeping their catch alive, the specimens were relegated to formaldehyde for later study by Smith-Vaniz.

Later that year, John E. McCoster, director of the Steinhart Aquarium in San Francisco, brought back the first live specimens of the Caribbean flashlight fish and placed them on display.

With the exception of how they masked their light organs, probably the biggest difference between the New World and the Old World species of flashlight fish is their sociability. The New World species seem to be shy, reclusive loners that abhor daylight and seldom approach closer than one hundred feet to the surface at night. Their counterparts on the other side of the world—both *Photoblepharon* and *Anomalops*—are so gregarious that their schools may number over two hundred individuals and are not shy about approaching the surface. McCoster reported that, along the coast of the Sinai Peninsula during the Six Day War, Israeli night patrols spotted a large green glow near shore. Suspecting an attack by enemy frogmen, they tossed grenades at the bright spot. Next day, everyone was surprised to find the beaches littered with the bodies of small dead fish whose eye pouches continued to glow green.

If flashlight fish use their light organs for luring prey, avoiding predators, and improving their vision, could the blinking lights also serve as a form of communication, wondered ichthyologist Eugenie Clark.

To see what response she might get, Clark carried a hand-mirror underwater with her on a night dive to see how *Photoblepharon* reacted to his own reflection. Singling out a fish, Clark held out her mirror to it. The fish responded by leaving the school it was in and following its own mirrored blinking image for some distance down the reef. Was it her imagination or had the fish increased its rate of blinking while watching its reflected image, questioned the ichthyologist.

Later, she asked Harvard University bioluminescence authority J. Woodland Hastings about *Photoblepharon*'s blinking pattern being a means of communication.

"We don't know whether blinking represents an actual 'language,'" answered Hastings, "[but] blinking rates change when two fish meet, or when one sees its mirror image as you discovered on your dive."

Hastings said that they knew the blink rate of the female of the species changed when she defended her personal territory and that the rate changed radically when a fish was disturbed. But whether this represented communication remained a question.

To learn more about this possibility investigators at the Hebrew University at Elat are putting the flashlight fish's blinking ability to the test with a dummy flashlight fish ingeniously rigged with blinkers controllable by the investigators. The idea is to vary the signals of the surrogate model and record the blinked responses from live specimens. Final results of this research should prove interesting.

Now that scientists can study the differences between all four family members of flashlight fish, they may learn why these differences occurred. For example, if the two New World species are employing both the rotating light organ and upward occluding lid, both mechanisms were probably present in their common ancestor. This hypothesis suggests that the flashlight fish's ancestor was present before one million to three million years ago. Scientists believe this is when the Central American land bridge that separated the aquatic populations of the Caribbean and eastern Pacific formed, thereby giving rise to the differences and separate species of the same family. But for those on the same side of the land bridge, why did the occluding lid mechanism of *Photoblepharon* become dominant in that species rather than the rotating light organ of *Anomalops*? Are the symbiotic bacteria in the light organs of all four species alike and interchangeable? Where do these bioluminescent bacteria come from in the first place? Are they passed from parent to offspring, or are they acquired elsewhere? These and other secrets the hidden lanterns have yet to reveal to us.

4

~~~~~~

Solving the Secrets of Sharks

Sharks have always been silent, feared predators of the sea, noted for their supposed unpredictability. Even today, after years of study by scientists using the most sophisticated electronic instrumentation coupled with the most modern technology of our age, the shark is only now beginning gradually to reveal its secrets. The more we learn, the more we marvel at this incredibly well-made animal; it is designed so perfectly that nature has had to change relatively little in the original model, which has been around for the last 150 million years.

Unlike the noisemaking fish or vocalizing marine mammals, sharks utter no sound, neither in the lower frequencies we might hear nor in the ultrahigh frequencies of the echolocating dolphins and whales. How strange that throughout millions of years nature has not seen fit to let sharks so much as croak at one another. Throughout time they have remained mute.

But you can be sure that, in that same length of time, these animals have developed other faculties. They may not say anything, but they certainly listen a lot. They are supersnoops, the oceans' most finely tuned, ultrasensitive eavesdroppers. What they are capable of detecting and how they use this information is astonishing. It is a one-way kind of communication designed to benefit that one recipient, the shark.

From the tip of its snout to the trim tab of its caudal fin, the shark is a mass of sensory receptors. In one form or another its body is covered with them. The animal is a living computer that moves steadily through the oceans, continually sensing its environment,

71

Black arrows point to a few of the many sensory pores on a shark's head. Note the regular line of pores to the right of the jaws. These are part of the shark's lateralis system that records mechanical changes in the water (pressure differentials). The more numerous and widely dispersed pores are openings to the ampullae of Lorenzini, the sensory receptors responsible for the shark's reactions to weak electrical fields enabling him to locate prey that may be hidden or otherwise invisible. These systems provide the animals with their unique sixth sense that picks up silent signals from the sea.

picking up sights, sounds, scents, electrical and magnetic disturbances with some of nature's most sophisticated sensing equipment. The animal processes all this information, running the printout through its primitive cerebral message center, and then reacts accordingly.

Possibly 95% of the information that sharks collect in this manner and process is information relative to perpetuating themselves in the same way as they have for millions of years.

While sharks are extremely well endowed with sensory equipment, experts knowledgeable in such matters have never given the animal high marks for intelligence. Some scientists rate their intelligence no higher than that of a dog, while others, less generous, say they have an IQ equal to that of a rat.

In 1958, when ichthyologist Eugenie Clark asked animal psychologist Lester Aronson if anyone had ever made a study of the learning behavior of sharks, he told her that, with the exception of a few basic olfactory experiments on dogfish, nothing very sophisticated had ever been tried on sharks since they were thought to have poor eyesight, poor mental capacities, and in general were considered poor subjects for the kind of experiments that had been performed with the higher animals.

As the marine laboratory begun by Eugenie Clark evolved into the Mote Marine Laboratory near Sarasota, Florida, it became a center of shark research for scientists throughout the world. This aerial view shows laboratory and shark pens where more recently scientists tried training a dolphin to attack sharks. The dolphin was to work with and protect underwater researchers from possible shark attacks. Courtesy Mote Marine Laboratory.

At that time, however, Clark had an excellent opportunity to try out any ideas she had in this area of research since she was the director of the Cape Haze Marine Laboratory on the Gulf of Mexico near Placida, Florida. She had started the lab three years earlier at the invitation of Ann and William H. Vanderbilt as "a place here where people can learn more about the sea."

In a modest twelve-by-twenty-foot laboratory building, she began collecting and studying fish found in the nearby waters of the Gulf Coast.

Soon, other scientists began appealing for assistance in their various research programs. For example, Dr. John H. Heller, director of the New England Institute for Medical Research, wondered if Clark could provide him with shark livers for his cancer research. Since Clark had been doing some studies of her own on the mysterious abdominal pores of some sharks and other primitive fish, she and a local retired commercial fisherman went into the shark-fishing business.

Dr. Heller and his wife came and stayed to carry on his research on sharks. Other scientists soon followed to pursue their particular programs of interest. Meanwhile, Clark involved herself in all phases of the activity, diving offshore with Aqualung to acquaint herself with the fish population of that coast and to add to her collection of specimens, while studying the ever-fascinating shark. She found it not only an intriguing animal, but also one easily obtained practically at her doorstep.

Before long, the laboratory's large pool contained vigorous specimens of lemon, dusky, tiger, and nurse sharks. Now, she wondered how one might go about undertaking a kind of Pavlovian conditioning program to find out what basic responses could be produced in some of her charges.

When this question was put to animal psychologist Aronson, he suggested training a shark to take its meal—a mullet—suspended in front of a sixteen-inch, square, white plywood target. To get the mullet, the shark would unavoidably press against the target, which would ring a bell that it could hear underwater. The idea was to condition the shark to associate the target with food and to press the target ringing the bell even when food was not present.

Training began with a pair of nine-foot-long male and female lemon sharks (*Negapion brevivostris*) that soon learned to take a

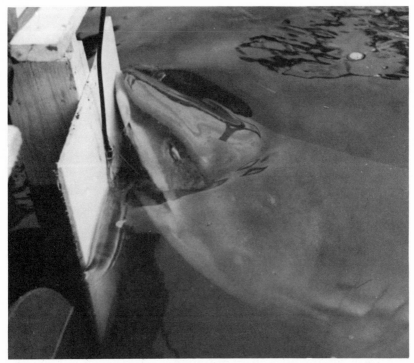

When ichthyologist Eugenie Clark wondered if sharks could be conditioned to perform on cue for food rewards, it led to a unique pioneer study of the learning behavior of sharks. Here, a nine-foot-long lemon shark simultaneously takes food and presses square white plywood target to ring bell. It soon was conditioned to perform more complicated feats revelatory of its learning ability. Photo courtesy William M. Stephens.

mullet dangled in front of the underwater target and simultaneously push the target, ringing the bell in the process.

With this feeding method Clark was surprised to find that it took only fifteen pounds of mullet a week to keep the large lemons healthy and apparently content in captivity. If, however, she fed them as much as they took, each shark consumed about thirty pounds of fish at a time, then refused food for several days afterward.

This seemed to indicate that possibly sharks were not the voracious and constantly hungry creatures man had long believed them to be. While there might surely be a difference between the appetites of captive and free-roaming sharks, her observation seems to be quite valid. There is at least one reported incident in an Australian aquarium in which a shark survived in captivity without food for a full year.

Another long-standing myth—that sharks must roll on their sides to feed—was repeatedly disproved as Clark watched the pair of lemons daily nose their targets, ringing their own dinner bell before they fed, head-on, without rolling.

Finally, after six weeks of such training, Clark decided to see if her students had learned anything. On the day of the big test, at the sharks' usual dining hour, the white target was presented underwater without any food. The large male lemon shark, usually the first to feed, rushed up to the target with jaws agape, looked it over, found no food, and turned aside. Then, it made eight more passes at the target, each time searching its white, sixteen-inch, square surface for its usual lunchtime mullet. Finally, when none appeared, the shark made another pass at the target, grazing it enough to ring the bell.

Immediately, Clark tossed its mullet. After that, the shark returned repeatedly to the target, nudged it with its nose, ringing the bell and receiving its reward. Both lemons were soon conditioned to ring the bell for their food.

Now began the next phase of training. Clark wondered how she could condition the sharks to change their swimming patterns and test their abilities to learn how to cope with a new set of circumstances.

The sharks habitually swam in a clockwise pattern about their pool. Now, when they pressed the dinner-bell target, their food was thrown further from the target. To reach it they had to turn left instead of right and swim in an unfamiliar pattern, reaching the mullet within ten seconds or it would be withdrawn from the water. After a few mistakes for which they missed getting their meal, the sharks soon learned to make the counterclockwise turn and reach their food within the ten-second limit.

When her two charges learned to rush up to the target, push it with their snouts, then swim swiftly to a spot some distance away to catch their mullet reward, Clark was elated. A dolphin or a seal could perform this simple maneuver almost instinctively; she was delighted to see that the lowly shark could also be trained to do it.

Clark felt sure that the sharks were hearing the ringing of the bell underwater and that this stimulus was part of their learned pattern, which resulted in their being rewarded. If this were true then, what would happen if the stimulus failed to occur, if the bell did not ring when the shark pushed the target?

To find out, she disconnected the bell. On the next feeding session the large male lemon charged up to the target, pushed it and turned counterclockwise to go to the feeding area for its food reward, which Clark dropped. But the shark never reached the food. Instead, it stopped midway, whirled around and returned to push the target again. Then it went to the feeding area and collected its reward. From then on, the shark repeated this pattern of pushing the mute target before going to recover its reward.

Do certain colors and objects of high reflectivity trigger shark attacks? Scientists say they do. When the U.S. Navy tried different colored sacks in testing C. Scott Johnson's "Shark Screen," sharks attacked yellow- or silver-colored sacks, but avoided those colored black. Here, Johnson tests the plastic survival sack while a shark ignores its presence. Photo courtesy U.S. Navy.

For over a year, Clark and her assistants worked with the lemon sharks in these conditioning experiments. Steadily the animals had learned each new step of the training program presented to them. Now, out of curiosity, Clark wanted to try a slightly different test. Most of the academic shark literature up to that time stated that, since sharks possessed no cones in the retinas of their eyes, they therefore did not perceive color as did creatures that had cones.

But was it true? Were sharks really color-blind? This pioneer schoolmistress to sharks decided she had an excellent opportunity to find out. Clark changed the color of the food target from white to a yellow panel, suspecting that this change would not really effect her conditioned lemon sharks. But she was wrong. And she was unprepared for the resulting tragedy.

At the sharks' customary feeding time the yellow-painted target was immersed in the pool. From the far end of the shark pen the male lemon made his customary run toward the target. Just in front of it, he abruptly stopped, reared backward, hurling his nine-foot-long body completely out of the water, and streaked to the other end of the pool. Suddenly, all the sharks in the pool were agitated, swimming in swift, erratic circles around the pen, moving first fast, then slow, bumping into each other as if totally disoriented.

From that moment on, the male lemon shark never recovered from the trauma. It stopped eating completely and swam in a peculiar manner with his body twisted, his head held to one side. Had he somehow sprained his back in the violent backflip? Or had the shock of interrupting his conditioned reflex been so great as to cause this kind of response? Was it perhaps indeed a combination of the two factors?

Clark never learned the answer. But the shark continued to refuse food and to swim erratically, always close to the surface, his back arched abnormally, the upper lobe of his caudal fin lolling uselessly on the surface. Sometimes he was seen to straighten out his body, but these moments were rare.

After his death, Clark dissected the shark to see if she could learn what had happened. But all she found out was that the lemon's liver was shriveled and leathery, indicating that it had, in all probability, starved itself to death.

The unfortunate death of an animal she had worked with for over a year and which had accomplished so much in its training did not deter the ichthyologist from pursuing a new approach to her experiment. Why had the color change provoked such a violent reaction in the lemon shark, she wondered. In an effort to find out, she tried targets of different sizes, shapes, and colors, introducing the changes gradually to her charges.

There was no repetition of the violent reaction experienced by the lemon. Yellow did not prove to be a shark-repelling color. Indeed, in some cases it was found to attract sharks. It was not just accidental that certain pieces of scuba diving equipment were said to be colored "yum-yum yellow" due to this attraction. But what no one was able to learn was whether this was due to the actual color yellow, or simply to the fact that the brightness attracted sharks. In subsequent tests, Clark found that many sharks responded better to the yellow target than to the white and seemed to learn faster. In analyzing the dramatic change that had come over her lemon shark during the traumatic test, Clark decided that the shark's sudden neurotic behavior had been due to the abrupt change in its already-established and highly trained routine. She wondered if this could be a clue to how sharks might be repelled by something that would short-circuit their psychological responses, causing them to retreat instantly and permanently lose their appetite.

Manufacturers of scuba-diving equipment as well as the U.S. Navy were extremely interested in knowing whether sharks were attracted or repelled by certain colors. Clark tried to carry out some color-differentiation experiments with lemon sharks, but found it to be an extremely complex operation. It required considerable prior knowledge about a shark's ability to respond to different wavelengths of the color spectrum under a wide range of uncontrollable variables such as the clarity of the water and the intensity of available light—so many, in fact, that it proved useless to pursue this course of research under the lab's limited facilities.

She did, however, learn some interesting things about a shark's ability to differentiate between different target shapes. She found, for example, that one of her conditioned lemon sharks had trouble telling a square from a circle when these shapes were presented as targets. Why, wondered Clark. Was it possible that the more primitive eyes of the sharks were seeing these targets differently than we might see them? Were the receptor rods in a shark's eyes registering only part of a target in such a way that the square and the circle

During her research with sharks at her Cape Haze Marine Laboratory, Eugenie Clark is shown working with one of her charges, an eight-foot-long
anesthetized bull shark. Photo courtesy William M. Stephens.

appeared alike? If that was the case, Clark reasoned, then she needed to test the shark's reaction to two targets of equal area and reflectability, but shaped differently in such a way that the shark saw them as two separate targets. The two shapes that best fulfilled these requirements were two identical white squares, the only difference being that one would be presented as a square while the other would be tilted forty-five degrees and presented as a diamond.

When these targets were shown to her conditioned lemon shark in subsequent tests, the shark unerringly selected the square to press for food, never once touching the white diamond. It zeroed in on the white square eighteen times in five testing sessions. Next, Clark sought to see if her sharks could differentiate between striped and unstriped objects in the environment. Pilot fish that habitually travel with pelagic sharks have vertical black stripes. Would this fact make these sharks and others like them more likely to accept vertically striped objects over horizontally striped objects, or would the sharks even be able to tell the difference between the two?

Clark's tests revealed that the lemon shark was quite capable of telling the difference between a plain white square and one with vertical black stripes an inch wide. But was the shark attracted to the white, unmarked target because it was brighter than the other, or had it actually recognized the vertical black stripes?

Before she could learn if the shark responded to a vertically striped target, as compared to a horizontally striped target of the same size, an abnormal tide in the area raised the water higher than usual, flooding the shark pen and her beautifully conditioned test shark escaped. Other circumstances, including moving the laboratory to a larger, better-equipped facility at Siesta Key near Sarasota, prevented her from continuing with the conditioning experiments.

By 1963, two young student experimenters at the laboratory successfully conditioned a baby nurse shark to discriminate between vertically and horizontally striped targets. Clark was surprised to learn that, indeed, this was quite an easy differentiation for sharks to make, not only in the lethargic nurse sharks, but also in the more active members of the clan.

Despite the fact that Clark's subjects were incapable of uttering a sound, their responses to her tests told her things that were previously unknown about these remarkable animals. Soon, other researchers were turning to the silent predators in the hope of learning even more intriguing things about their behavior. They wanted to know, for example, more about the shark's unique ability to home in on underwater sounds. Hans Hass had been aware of this phenomenon from his underwater observations of the shark's response to speared fish off the Curaçao coast in 1939. There, in water with visibility of over 130 feet, whenever Hass speared a large fish, a shark would appear within 20 seconds. Having had to come from a distance of at least 140 to 1,000 feet away, the shark had not had time to pick up the scent of the blood in the water. Something else had attracted it. Was it the wriggling of the speared fish that they had heard? Hass decided it had to be this. He believed that the sharks were picking up the sounds of the struggling fish with the sensory cells of their lateral lines, then quickly homing in on the potential meal.

By the end of World War II, Hass had the idea that, if he could record the sound of struggling fish underwater and this sound attracted sharks, then this method could be used by commercial shark hunters to great advantage. Moreover, it might prove to be a way to get rid of the menace to bathers in shark-infested waters such as on the Australian and African coasts. So Hass headed for the Red Sea to dive with sharks and see if he could record some of the sounds of speared fish on a special underwater microphone called a hydrophone.

Some time after that, I saw the results of those early efforts in a Chicago release of Hass's famous underwater documentary *Under the Red Sea*. The film was filled with footage of the most remarkable kind. Remember, this was several years before audiences had a chance to see Jacques Cousteau's marvelous *Silent World*.

Two incredible scenes in the Hass film will always be remembered. In one, three-quarters of the theater's screen was blotted out by the enormous shape of a great whale shark's head while the diminutive figure of a man swam up to the monster's enormous maw, looking like a mosquito buzzing around an elephant. In the other scene, Hass had deployed his special underwater loudspeaker (transducer) and withdrawn to film a local fish population's reaction to extraneous noise played over the loudspeaker, filling the water with sounds of gunshots, sirens, and traffic racket. As his camera recorded the event, the fish paid not the slightest attention to this cacophony. But, according to the film's narrator, when Hass played a Strauss waltz over his underwater speaker, the fish began coming by twos and threes from everywhere, joining together in a kind of swimming waltz around and around the speaker until that instrument disappeared from view behind a wall of fish. I can remember how impressive this scene was to me, as it was apparently to the film editor who put the Hass documentary together.

The real truth about that scene was not revealed until years later when Hass himself explained it. Apparently during one of Hass's underwater photographic sessions, his topside helper became bored and broadcast a jazz record over the underwater speaker. The music came through loud and clear, but the fish paid no attention to it.

The next record was a Viennese waltz. The fish were still unimpressed. While watching them, Hass saw a large school of no less than three hundred big silver jacks approach in tight formation and begin circling him and the loudspeaker. The fish stayed about ten feet from the speaker and almost seemed to be moving in rhythm to Johann Strauss's waltz. Possibly they were attracted by the music, but surely they were not performing because of it. Still, Hass left the scene in his film because their movement coincided by chance so perfectly with the waltz tempo.

"When an American version of our film was subsequently produced in Hollywood," said Hass, "the American film editor placed the waltz at the center of the action. Our experiments with inaudible water vibrations did not impress him much, and so he made us emit all kinds of strange noises into the sea. Only when cow bells, revolver shots, children's crying, and other sounds had failed to make any impression on the fish, did I—in this version of the film—conceive the brilliant idea of trying a Viennese waltz. And now, the fish grouped themselves into pairs—the editor used for this purpose the shots we had made of fish in love play—and soon everything was swimming and dancing to the waltz tune. We had reserved the right to reject alterations during production, but we were only shown the version just before the New York premier, when it was all finished. There was nothing left to us but to point out the error in interviews and to accept with good grace the congratulations heaped upon us for proving that fish were musical creatures."

Hass failed in his efforts to record the underwater sounds of fish that attracted sharks. He was operating under especially trying conditions. In the Red Sea heat, the edges of his magnetic tapes melted and his underwater recordings seemed to have picked up more interfering noises, including the clanking of the microphone cable across the floor of the ocean, than it did the fin beats of the fish. Later he tried it again under somewhat better conditions in the West Indies, but for various reasons was once more unsuccessful.

Meanwhile, however, Warren J. Wisby and his graduate students Donald R. Nelson and Samuel H. Gruber at the University of Miami's Institute of Marine Science accomplished what Hass had tried to do. Wisby, too, had long believed that sharks were using sound to home in on animals in distress. He remembered that many skin divers had reported that they never saw sharks until after a fish was speared. Then, the sharks arrived, oddly enough, often from upcurrent, a direction from which they could not have smelled the fish or its blood. This seemed to substantiate Wisby's belief that the sharks were being attracted by other factors. Again, with the same reasoning that Hass had applied, Wisby felt that if he could record this particular kind of sound, it would be a method of calling sharks. A further projection of that idea might have included the possibility of attracting these creatures away from populated beach areas.

Wisby was well aware of the experiments of George H. Parker of Harvard University, who some sixty years earlier had found that sharks deprived of both sight and hearing could still respond to a source of disturbance in the water if their lateral line systems were intact. When the main trunk nerves of this system were severed, the shark failed to react to a disturbance. But nobody knew how sensitive this system was or over what distances sharks could respond to disturbances until Wisby's experiments in 1961.

Wisby first decided to determine what kind of sounds were made by struggling fish. With the assistance of Nelson and Gruber, they boated out to the reefs off Miami with tape recorder, hydrophone, and scuba gear. Diving to the reef, they located a large black grouper in a cave. One of the men speared the fish while the other held the hydrophone that monitored its struggles, which were in turn recorded on the surface by Wisby.

Back at the laboratory, the tape was analyzed. Along with the low-frequency sounds of the struggling fish were the sounds of the spear shaft clattering against coral rocks. The researchers recorded each sound individually on other tapes until they had identified three forms: low-frequency pulsed or interrupted sound, high-frequency pulsed sound, and low-frequency continuous sound.

Next, they returned to the reef, hung an underwater transducer just above the bottom, and began playing back their various tapes at various volumes while a diver watched the response from above. The low-frequency continuous sound produced no results. The high-frequency pulsed sound attracted two sharks. But when the researcher played back the low-frequency pulsed sound—the same sound made by struggling fish or a thrashing swimmer—the results were astonishing. Over a nine-day period, eighteen sharks up to nine feet long—including hammerhead, bull, lemon, and tiger sharks—were attracted to the transducer. They approached, swam warily around the speaker searching for the sound source. Finding no food, they veered off and circled the area at a distance.

Further work and analysis of the results produced these interesting facts: sharks hear in a frequency range of from 7.5 to 400 cycles per second (CPS), as opposed to the human hearing range of from 40 CPS to 2,000 CPS. But apparently not all sounds attract sharks, only those occurring in the lower part of their hearing range, from 7.5 to 100 CPS—their hunting range. This meant that at the lower end of the scale they were able to hear sounds that we could not.

Struggling fish or humans were actually broadcasting bursts of these low-frequency sounds that sharks were able to pick up as much as six hundred feet from its source. Since these bursts of sound were traveling through water at the rate of a mile per second, a shark's appearance on the scene seconds after this "dinner bell" had rung, was now much easier to understand.

In the years that followed, Nelson, Gruber, and others went on to investigate the remarkably developed sensory systems that make these animals unique. Samuel Gruber wondered just how much a shark actually saw in this strange underwater language of sights, senses, and sounds. Like Eugenie Clark, he was curious to know whether or not sharks could actually see colors, or if they differentiated solely on the basis of light reflectivity of a subject. In other words, were they able to detect the difference by seeing things in different shades of gray? Also, how good was their eyesight for seeing in the dark or near dark?

These were the questions Gruber sought to answer when he set up an ingenious device at the Miami Marine Institute for studying sharks' vision. An experimental shark was secured to a platform in a tank of circulating water. One end of the tank was a Plexiglas bubble into which the shark's head was fitted. Outside this window a researcher flashed a small filtered light of various colors and intensity. Each time the light flashed, a mild electric shock caused the shark to blink. This was repeated until the shark became so conditioned that it blinked whenever the light flashed, indicating that it could see the color and intensity of the light being used.

With this equipment, Gruber and his assistants soon learned that most sharks indeed perceive color and are fully capable of seeing in extremely low light. Moreover, analysis of their eye structure revealed that they can apparently tell the difference between an object and its background in dim light, a faculty that makes them highly efficient hunters under adverse water conditions.

Despite this knowledge, there is still mixed opinion among authorities on just which colors sharks prefer. Here are some examples that have helped confuse the issue. In 1958, fishing in shark-infested waters off Clipperton Island, an uninhabited isolated atoll six hundred miles southwest of Mexico, the nearest land, the late Conrad Limbaugh found that when shark bait was concealed in a fluorescent orange or a white sack, the local species, largely Galapagos (*Carcharhinus galaeagensis*) sharks, struck the bait concealed in the white sacks almost three times more than they did the bait in the fluorescent orange sacks. This seemed to signify that this species, at least, was attracted more to the white concealed bait than to the other. Could it provide a clue to all sharks' color preference?

This has long been a controversy. Pioneer shark hunter Captain William E. Young, who spent thirty years of his life catching sharks in the early 1900s, firmly believed that anything white attracted them. Sometimes he found that newspapers spread on the ocean's surface "chummed" the sharks within range. When netting sharks in Pindimar, Australia, he wrote: "We decided to use nets, the like of which had never been seen in Australia before. They were about 1,000 feet long, 16 feet deep, with an 8-inch mesh. They were hung in the same 'curtain' fashion that I had found to be so effective elsewhere. In my entire sharking career never had I been given an opportunity to test out the theory that sharks were attracted to white or light objects. I decided to try an experiment with the nets. We alternated blue, green and white sections in one net. Invariably, we found sharks in the white section—and none at all in the colored. The experiment left no doubts, in my mind at least, about the effect of color on a shark's senses."

And then there are the amazing *ama*, the pearl-diving women of Japan who purposely wear flowing white garb when they dive deep for pearl oysters because they have long believed that white *repels* sharks.

In 1971, when E. B. McFadden and C. S. Johnson tested the color and reflectivity of large plastic bags with flotation collars intended to shield survivors in the sea from sharks, they found that the yellow-painted survival gear was attractive to free-ranging sharks while the same gear painted black was apparently ignored by them. When I asked Johnson if they had tried silver, which is the color of water's undersurface, he said, "Yes, they came in and took a bite out of it!"

Later, when Samuel Gruber acoustically attracted silky sharks (*C. Falciformis*) to three polyethylene globes forty centimeters in diameter separated by six meters—one globe painted fluorescent orange, another black, and a third white—the sharks clearly avoided the fluorescent orange globe, but readily removed bait from the black globe, and less frequently from the white globe. You can see how sharks got a reputation for being unpredictable!

Some investigators wonder if this kind of research into the visual orientation of sharks is worth pursuing. In most cases the only thing everyone agrees upon is that objects of bright colors and high reflectivity are often more attractive to sharks than dark objects.

Where once it was believed that a shark's brain was almost totally devoted to the sense of smell, researchers were to find that only about 10% of the forepart of the shark's brain actually is. The remaining 90% is most likely involved with other sensory input. Scientists have found that the mental makeup of sharks is far more complex than previously imagined. Certain sensory cells located on the shark's heads are chemoreceptors, capable of recording the minutest changes in the chemistry of the water. Sharks "taste" with these cells. There is now evidence to indicate that they are able to detect chemical changes in the water brought about by wounded animals. There is even one school of thought that humans under stress may exude a scent of fear that triggers a shark attack.

Controlled experiments revealed that combinations of amines

How do the fish swimming near this shark know they are not in immediate danger of being devoured by the shark? Scientists believe the shark's swimming manner communicates this information to the food fish. This shark is cruising, not feeding. Photo courtesy Marineland of Florida.

and amino acids in the water caused sharks to go into feeding frenzies, biting everything within reach, including floating debris and surface bubbles. Is this response to the amino-acid combination the reason sharks go into a frenzied behavior whenever they are exposed to water where blood or other body juices of animals are present? Apparently, this kind of scent or taste does indeed trigger a feeding frenzy. Observers have seen injured sharks suddenly grow even more stimulated when they swam through their own scent trails.

In related experiments, Edward S. Hodgson and Robert F.

Mathewson set up an underwater television camera at a depth of sixty-five feet at the edge of the Gulf Stream near Bimini to see what effect an amino-acid mixture would have in enticing wild sharks up from the ocean's depths. It was an ideal arrangement, enabling the unobtrusive study of a shark's behavior under this stimulus by observers monitoring the whole operation on closed-circuit television from the ocean's surface.

The source of the scent, a bottle of amino-acid mixture placed within a concrete block, leaked its essence into the sea, to be carried downcurrent. In a short while, surface observers saw a variety of fish swim into the area, including the larger types such as the Nassau grouper. All seemed inordinately interested in the area around the sensors. Interestingly, *most of these fish cleared out of the area as if by signal at least thirty to ninety seconds before the first sharks appeared*. (Had they *heard* the distinct and meaningful shark-feeding swimming sound?) In repeated tests this occurred so punctually that it became a signal to observers that sharks would soon appear on the scene.

When they did, the scientists saw lemon sharks, nurse sharks, and sharp-nosed sharks *(Rhizoprionodon terraenovae)* move into the area. Knowing the direction of the current, the researchers could tell whether the sharks approached from downcurrent or upcurrent. The lemon sharks invariably approached by moving upcurrent; they did not remain in the area, but paused momentarily and then moved on. Both the nurse sharks and the sharp-nosed sharks approached from various directions, and in twelve out of seventeen positive identifications, they arrived from a point downcurrent.

It was therefore concluded that these animals were indeed following the invisible scent trail being distributed in the water, but unlike the lemon sharks, the nurse and sharp-nosed sharks refused to leave the site of the appetizing scent. Instead, after circling the cement block protecting the bottle several times, some settled down to lie immobile with their snouts pressed against the block.

The televised scene was not clear enough for the scientists to tell whether or not these sharks were vicariously enjoying the appetizing scent by jaw movements that might anticipate a forthcoming feast; but in other experiments, this often occured. In one case a mixture of TMAO and glycine, the breakdown products found in tissues of excreta of fish, summoned a nurse shark from beyond a break in the reef to the plastic source jug and apparently proved too appetizing for the animal to resist. It chewed up the jug.

All of this indicates to scientists that these animals could quickly pick up a scent carried by the current and direct themselves to its source by following its highly diluted trail. Literally, they followed their noses. In view of the almost magnetic attractions of some scents to these marine creatures, one cannot help but wonder why researchers have encountered such great difficulty in finding a reverse stimulus, scent in the water that repels sharks.

During World War II, scientists thought they had found the answer when they learned that most sharks were repelled by the deteriorating flesh of their own kind. Analyzing the constituents of deteriorating shark meat chemists isolated the probable byproduct of the tissue breakdown—copper acetate. Subsequent tests proved that this chemical indeed caused sharks to flee. Here was the magic substance that seemed sure to provide the answer for a bonafide shark repellent. Copper acetate was combined with a nigrosine dye to serve as a marker, packaged and distributed to our airborne and oceangoing troops as "Shark Chaser."

As Perry Gilbert and other scientists who researched this project during World War II and after were to learn, the compound called Shark Chaser was actually little more than a psychological aid, a crutch for downed aviators or torpedoed seamen with no other defense against shark attacks. No one really knows how many lives were actually saved by Shark Chaser because, instead of giving up and letting the sharks have their way, the menaced men made that last-ditch effort to survive, simply because they felt secure in having what they believed was a potent weapon against the sharks. But as Gilbert and the others now know, Shark Chaser was an otherwise impotent defense against sharks. Tests after the war showed that sharks in a feeding frenzy would actually swim into the heavily dyed water and gulp down packs of Shark Chaser!

Thus, scientists have been discouraged in their long search to find a satisfactory chemical shark repellent. They say the reason that such repellents lose their effectiveness is because they are rapidly diluted and dispersed in open water. Yet, certain substances such as that exuded by the Red Sea Moses sole (*Pardachirus marmoratus*), a flatfish almost identical in appearance to the common flounder, are apparently such potent shark repellents that Eugenie Clark has recorded instances in which the Moses sole has, with its toxin, stopped in mid-bite the jaws of a shark as it attacked. To date, this poison is probably the most potent shark and barracuda repellent known to man. The milky substance is produced freely from 240 poison glands located along the dorsal and anal fins of the sole. Tests have been run on the substance since 1972. There is no doubt as to its effectiveness as a repellent. When Clark strung up a wide variety of live and dead foodfish for sharks, including several dead and live Moses soles, and left the line in shark waters overnight, only the soles were untouched by the sharks; all the other bait was gone. If alcohol is used to remove the toxin from the sole, a shark will immediately gulp down the flatfish. So, the repellent's effectiveness seems to be that it is constantly being deployed into the water, which may make it impractical for our use. But research is continuing in this country on all these possibilities. One can imagine a time when swimmers in a high-risk area might anoint themselves with synthetic Moses sole shark-repellent oil and actually swim among sharks without fear of attack.

Meanwhile, other scientists, such as Barry L. Roberts of the Marine Biological Association Laboratory at Plymouth, England, and Arthur A. Myrberg, Jr. at the University of Miami's Rosenstiel School of Marine and Atmospheric Science, continue to expand our knowledge about the shark's lateralis system and other receptors it uses to tune in on the kind of sounds that attract them.

Most of the early work in this area was pioneered by G. H. Parker, who in 1905 recognized and accurately understood the functions and significance of these sensory cells. Parker's discovery that certain behavior could be triggered in fish by a tuning fork, causing it to perform seemingly miraculous feats of sensory perception, has provoked a considerable amount of discussion among scientists over the last seventy-five years. Everyone wonders how large a role this unique system—this lateral line—plays in the natural behavior of marine animals. Some scientists believe the lateralis system is sound sensitive. Others, strongly opposed to this view, say the system is really sensitive to water displacement, the compression and noncompression of water molecules. Apparently, these disturbances are created by moving objects in the water. When a fish reads these disturbances through its lateral line, it can recognize its friends, its mates, and its enemies. The system may be used to help the fish navigate through echolocation; it also may enable the animal to school in orderly fashion. The question is, how great a use is it to sharks?

In the 1940s, not long after man entered the sea as a skin and scuba diver, he began to learn that sharks and sound were somehow related. But here again, the open-ocean encounters with these creatures often had opposite results. After his encounters and experiences with sharks in the Red Sea, Hans Hass firmly believed that a loud shout made by a diver underwater would turn away an attacking shark. It had worked for him; he believed it might for others.

This advice, however, was regarded by other underwater pioneers, such as Jacques Cousteau, as "little short of criminal." In fact, Cousteau believed that sharks could be attracted in this manner. More probably they are both right, depending largely on the circumstances and the sharks involved.

The lateral line and its function has been an enigma since 1825, when it was known that the lateral line played some important sensory role, but nobody knew exactly what. It seemed to fall somewhere between a fish's sense of touch and its sense of hearing.

After Wisby, Nelson, Gruber, and Banner's investigations with sounds and sharks, Myrberg began extensively researching the qualities of these sounds and their effects upon sharks. His efforts soon revealed that not all carnivorous fish reacted in the same way to underwater sounds. Groupers and snappers, for example, are also attracted to the same kind of sound, though they are slower than sharks in responding to it. Sound, to be attractive to sharks, had to contain frequencies below eight hundred or one thousand hertz (Hz). Otherwise the sharks would not come to it.

Since the natural sounds that attracted sharks were seldom continuous tone sounds, these failed to interest them at all. The most attractive sound was the irregularly pulsed low-frequency type of emission. Although the researchers were able to create synthetically the type of sound that attracted sharks, their research seemed to indicate that most biological sounds—the natural low-frequency noise made by floundering fish or struggling swimmers—are detected by sharks at distances of less than three hundred feet from the source. But whether it was biological or synthetic sound that was being broadcast by an underwater loudspeaker, there was no need for other sensory stimulation to be present—such as chemical (blood or fish juices) or visual (food) stimuli—for the shark to be called in and immediately attack the sound source by striking, biting, and even swallowing the entire hydrophone.

Sound alone was the attractor. It was so powerful that observers noted it caused "hunching" in sharks, a characteristic kind of body attitude they adapt just moments before feeding.

As these sound researchers tried different frequencies of pulsed and unpulsed sound, increasing and decreasing their intensities underwater, they learned that, paradoxically, the same sound that attracts sharks can also repel them.

This response was reported by Banner in 1972 during his sound studies with young lemon sharks. Closely observing these test animals' responses provided Banner with clues to what caused the reactions. Banner noted that sharks sometimes took flight at the precise moment certain sounds were broadcast. This was especially true when sharks were approaching a source. But Banner also observed that rapidly pulsed sound seldom caused such a reaction. Could it be, he wondered, that during a shark's fast approach rapidly pulsed sound allowed the signal level to increase smoothly? But this would not occur with a single loud sound; perhaps only with one that was repeated after an interval of several seconds of silence. These sounds seemed to startle sharks into flight, in the same way that this reaction was reported by Hans Hass, who brought about the response simply by shouting loudly underwater. Scientists now call this the "scream theory" of shark repelling.

Said Myrberg of Banner's startle response reaction in sharks: "Strikingly similar withdrawal responses were first observed in silky sharks by our team in 1970 during a field study in the Tongue of the Ocean, Bahamas. Water depth was approximately 2,000 meters and the sharks were adults or subadults. . ."

Two years later, Myrberg more fully explored the idea that a shark homing in on a sound source that was entirely suited to it could be suddenly startled off when that same sound source abruptly changed. Thus, Myrberg was able to duplicate the same kind of attracting/repelling response simply by changing from a low-frequency pulsed sound to short intervals—two and one half seconds on, two and one half seconds off. That simple change in an accepted pattern was sufficient to bring about a strong flight reaction to sharks in the open ocean.

Was this the same thing Eugenie Clark had done when she

interrupted a conditioned response in her test lemon shark by presenting the shark with the unexpected, creating in effect a highly negative reaction? The similarity of these responses and how they were triggered are worth contemplating.

As Banner had observed, the startle response could be brought about during the smooth approach of a shark to the speaker by a sudden volume increase from fifteen to twenty decibels when the shark was just a few meters from the sound source. Was sound the secret to repelling sharks?

When Myrberg tried the Banner experiment on different species of sharks, he found they reacted differently to the change. Silky sharks, for example, took anywhere from ten to thirty seconds to vacate the area, while others responded within a couple of seconds of the signal change and veered off immediately, disappearing from view. He also observed that, by repeating these experiments and attempting to call and then repel the same sharks, after about six or seven repetitions of this the sharks apparently got over their fear of the signal change and failed to respond as they had. Repetition, it seems, had conditioned them. Concluded Myrberg: "Sound level per se cannot be the entire answer since one sound at a given level results in rapid withdrawal by sharks, while another at the same or slightly higher level, will attract them right up to the source. The real key to the problem may actually be the progressive increase in loudness as perceived by an approaching shark. . . . Some factor of intensity is central . . .[but] it is the manner whereby a given intensity is reached, i.e. rapidly or slowly, relative to some unknown reference."

This area still requires considerably more research before we will know more about how these animals use sound in their environment and how man can make use of this knowledge to make the seas safer for his presence there.

In 1917, researchers G. H. Parker and A. P. Van Heusen noticed a strange response in the North American catfish (*Ictalurus nebulosus*) in their laboratory. When they blindfolded their specimen, then reached into their aquarium with a glass rod, the fish failed to show any response until it was actually touched. However, when the scientists performed the same maneuver using a metal rod, the blindfolded fish avoided the rod as if it could actually see it. Why had the fish avoided the metal rod and not the glass? When blindfolded, how could it possibly have told the difference?

Such were the questions that confronted these early inquirers into the biological mysteries of fish. After successive experiments, the two men convincingly demonstrated that the fish were responding to the presence of electrical current generated between the metal rod and the aquarium water.

Though intrigued by this phenomenon and so close to discovering that their laboratory catfish possessed a unique sixth sense for detecting electric currents, neither Parker nor Van Heusen realized the full biological significance of this seemingly inexplicable behavior in their specimens.

Other researchers eventually picked up the trail of this strange phenomenon and noted its presence in other species. In 1934, S. Dijkgraas reported this odd sensitivity to metal in a certain species of bottom-dwelling shark common to the Mediterranean and coastal European waters. And twenty-five years later, A. J. Kalmijn confirmed this reaction and sought to learn more about it in sharks and rays.

Subsequently, Kalmijn, of the Woods Hole Oceanographic Institution in Woods Hole, Massachusetts, discovered that sharks, skates, and rays are remarkably sensitive to weak electrical fields. Further testing revealed that they are most receptive to frequencies ranging from zero to eight Hz direct current. They were able to detect this through sensory pores called the ampullae of Lorenzini that were systematically arranged around the protruding snouts of these animals.

To complicate things, the ampullae were associated with other skin sensors scattered over the sharks' snouts, sensors connected to the lateral line system. How could Kalmijn be absolutely sure which system was detecting the presence of electricity in the water?

By the process of elimination. Carefully and selectively he denerved all sensory cells other than the ampullae of Lorenzini, until finally he was able to demonstrate by laboratory test that these receptors alone were responsible for the sharks' reaction to weak electrical fields.

By testing the sharks and skates in his aquarium, Kalmijn found that his specimens produced direct current in a low-frequency voltage emanating from potential differences between their skin and water surfaces. Mucous membranes around the mouths and gills of his laboratory fish were found to produce steady direct-current fields. Since these fields were far stronger than the low voltages that sharks could sense, Kalmijn suspected that the sharks might be homing in on these stronger impulses with their electroreceptor cells.

To test this theory, Kalmijn performed some experiments. To see if a species of small shark, the spotted dogfish (*Scyliorhinus caniculus*), and skates (*Raja clavata*) could locate hidden prey as they might in their natural habitat, he released small specimens of the flounder (*Pleuronectes paltessa*) into a saltwater aquarium, and after the specimen had hid itself in the sand bottom, he added a few drops of whiting extract to the water to stimulate the appetite of the test animals. Almost immediately after they were placed in the aquarium, the sharks and rays began searching the habitat for something to eat. Cruising back and forth over the bottom of the pool, the test specimens approached to within ten to fifteen centimeters (four to six inches) of the hidden flounder, then made determined attacks on the quarry, routing it out of the sand and devouring it.

Had something other than electrical stimuli tipped off the test animals to the prey's hiding place? To find out, Kalmijn decided to enclose the flatfish in some kind of container that would allow transmission of an electrical current while preventing any other stimuli—e.g., visual, chemical, or mechanical—from escaping.

The substance he selected was an agar chamber made by dissolving 3 to 4% agar in saltwater to produce a rigid structure allowing free passage of electrical current while keeping the test fish otherwise concealed so that the shark could neither see nor smell the specimen.

As Kalmijn later reported: "The agar chamber was placed on the bottom of the pool, just under the surface of the sand. To keep the flounder alive, the chamber was ventilated with a steady flow of sea water. Despite these changes, the sharks and skates again made well-aimed attacks from the same distance and in the frenzied manner as when no agar was screening the prey. These results were in full accord with the assumption that sharks and skates can locate prey bioelectrically."

But did the one-centimeter-thick agar roof of the flounder's chamber really provide an adequate odor barrier? To be sure, Kalmijn substituted cut pieces of whiting, a favorite shark and skate food, for the live flounder, again burying this container of bait in the sandy floor of the pool. The sharks and skates were then introduced into the aquarium and searched fruitlessly, stimulated by the odor of the seawater ventilating the chamber. Still, the predators were unable to locate the food even when swimming directly over the chamber. And as a final test, Kalmijn put the flounder back in the chamber, covered it with a thin sheet of electrically insulated polyethylene plastic, and let the test sharks and skates back into the pool. Once again, though they searched the sand carefully for the prey, often passing directly over the polyethylene-insulated spot where it was concealed, they failed to find the flounder.

To establish even more supportive evidence that the test animals had been attracted electrically to the prey, Kalmijn simulated the flounder by creating a small electrical current equal to that generated by the prey animal using, instead of a flounder, two salt-bridge electrodes concealed in the sand. After being stimulated to search for food by their aroused sense of smell, the sharks and skates adopted the same characteristic hunting behavior toward the electrodes as they had toward the flatfish, attacking the source of the electrical current tenaciously over and over when they encountered the electrodes.

To see if his test would work with other species of sharks, Kalmijn repeated it with the lemon, the smooth dogfish, the leopard sharks, and several species of stingrays with the same results. Here, then, was concrete evidence of how these animals used their sixth or bioelectrical sense for hunting prey.

How would one go about observing this behavior with animals in their natural habitat, wondered Kalmijn. After all, scientific laboratories and fiberglass pools provided such conveniently controlled conditions. What would happen in the open ocean where conditions were far more complex and a prey's electric field would be mingling with those of chemical and physical stimuli? Moreover, how could an observer carry out such an experiment without inadvertently introducing into the test site extraneous electrical fields that would throw off the normal responses?

In 1976, Kalmijn learned that a species of shark called the smooth dogfish (*Mustelus canis*) nightly moved out of the depths into the shallow waters of Vineyard Sound to feed. Since this species was an

"Shark Screen," a black plastic sack attached to a floatation collar invented by C. Scott Johnson (shown testing the device) has proved to be an effective deterrent to shark attack, but is not yet in production. Photo courtesy U.S. Navy.

active bottom-feeder, Kalmijn decided that it might make a good test animal for his natural-habitat observations. He worked out a plan. It had to be a totally nonmetallic operation, for he was well aware of how Parker and Van Heusen had originally detected electrical sensitivity in their test catfish and sharks. Kalmijn did not want to spook off his animals by creating galvanic fields of his own.

Once they had worked out all the details, he and his student collaborators launched a rubber Zodiac boat one dark night and went to sea. Selecting a shallow-water site over a sandy bottom, they set up their apparatus. Essentially, they created a strong source of scent by piping small amounts of liquified herring through plastic tubing. The appetizing essence released into the water mass moved downcurrent. Then, twenty-five centimeters (ten inches) away from the odor source on each side of it were placed two pairs of salt-bridge electrodes to provide the electrical stimulus. Then, from the surface, the viewers watched their setup through a glass-bottom viewing box mounted on the stern of the boat over the bottom that was brightly illuminated by a one-hundred-watt underwater light.

Then the sharks came, sometimes singly, sometimes in groups of two to five, both young and adult members. It was apparent that they were searching the bottom, hunting for the source of the appetizing odor. But interestingly, when the sharks found it they did not bite it, but instead turned sharply away toward the set of electrodes that was emitting a weak electrical field. This they attacked vigorously, attempting to rip it apart. When the current was switched off from one pair of electrodes and the other set was activated, the sharks abandoned the dead electrodes and switched their attack to those newly activated.

Kalmijn concluded that odor cues in the water mass may attract sharks from great distances away, but when they approach the source, they then rely upon both visual and bioelectric sensors to find their prey. Said Kalmijn: "Our tests' results also indicate that attacks on humans and underwater gear may be elicited and guided by electric fields resembling those of regular prey. The human body, especially when the skin is damaged, creates in the water direct current bioelectric fields that sharks may detect from distances up to one and two meters. The galvanic fields of metallic objects on the body are often even stronger."

The fact that the Johnson Shark Screen polyvinyl bag and flotation collar effectively shields a potential victim visually, chemically, and bioelectrically from shark attack seems to make it our most effective protection against marauding sharks. A couple of years ago, after the U.S. Navy had made exhaustive tests on the Johnson Shark Screen, I asked C. Scott Johnson why the Navy had not made his protective device standard equipment. It seemed so superior to the still-used and ineffective Shark Chaser chemical repellent.

Johnson told me that as long as supplies of Shark Chaser repellent were available, the Navy did not plan to switch to the Shark Screen bag. Why, then, I asked, don't manufacturers, in the private sector, produce the bag for civilian use? I said that I thought it would be as practical to have a Johnson Shark Screen packet stowed aboard every airplane or boat for passenger use as it is to provide life preservers. Surely it could be put up in a very small package.

Johnson agreed that it would be a good idea, but said that most manufacturers were not interested in handling such an item until the Navy endorsed it and supplied it for their own use.

Granted, incidences of shark attacks on shipwrecked sailors or downed airmen occur rarely, but how comforting to be protected in the rare instances when it does occur. I know of at least one occasion when a family of two adults and two children spent the night drifting in the Gulf of Mexico after their boat sank in a storm. Sharks in the area did not molest them throughout the night and continued to keep their distance until a Coast Guard Rescue helicopter approached and began picking up the survivors from the air. Then, the sharks attacked and one of the youngsters lost his life.

Researchers believe there is a direct correlation between a shark attack and the pulsating low-frequency sounds of a helicopter's propellers, plus the water-surface disturbance caused by the aircraft's whirling blades. These combined stimuli appear to provoke a direct attack on survivors in the area. One wonders if such attacks could have been avoided if the survivors had some Johnson's Shark Screen bags. For anyone interested in fabricating their own, it perhaps is worth knowing that Johnson's first experimental model of the Shark Screen was made from a youngster's plastic wading pool, the kind with an inflatable collar around the sides. In an all-out emergency situation, it might be well to remember that the combination of a life ring and an extra-large, heavy-duty, dark-green polyethylene garbage bag might serve the same purpose.

Once he saw what a marvelous sensory system sharks, skates, and rays had for detecting the bioelectric fields of their prey, Kalmijn wondered what other underwater uses these animals might have for detecting bioelectric sources. Also, could they detect other invisible force fields known to be present in their watery environment? Was it possible for sharks, skates, and rays that possessed bioelectric sensors to induce electrical directional fields as they swam through the earth's magnetic field, which might then be used for open-ocean orientation and navigation?

This was not as farfetched as it might seem. To check its validity, Kalmijn tried some experiments. With Theodore H. Bullock, he placed a leopard shark *(Triakis semifasciata)* in an outdoor fiberglass pool at the Scripts Institution of Oceanography at La Jolla, California. With the test animal swimming around the circumference of this pool, the investigators placed a small induction coil beside the tank and set up a low magnetic field in the water. The force field was turned on when the animal was furthest from the induction coil on the far side of the pool where it could not possibly feel the effect. But as it continued swimming around the circular habitat until it reached the area opposite the coil, the shark suddenly turned out of the magnetic field and swam into the center of the pool. This experiment was repeated with several other leopard sharks and the response was always the same. The sharks apparently sensed the imposed force field even though it had not distorted the earth's magnetic field by more than 25%.

Next, the investigators noted that the sharks preferred resting in an area of the tank's circumference that was just a little off magnetic north. To ensure that the sharks could not visually orient themselves for this next experiment, Kalmijn and Bullock removed all structures from the pool, covered the twenty-one-foot-wide enclosure with a large sheet of black plastic, rotated the whole setup, and moved it to another site.

Surely, this should have thrown off the test sharks' sense of direction so they would choose a different resting area. But this was not the case. The next morning, when the researchers checked the animals, they were resting in the exact same sector of the pool, a little off magnetic north.

Had they been able to position themselves because their bioelectric sense told them exactly where they were in relationship to the earth's geomagnetic field? To find out, Kalmijn and Bullock, in effect, neutralized the earth's ambient magnetic field with two large induction coils mounted outside the tank. After that, the sharks apparently lost their sense of position, for they distributed themselves randomly around the pool.

Despite what these tests seem to signify, Kalmijn felt that neither was fully conclusive. The fact that the sharks had avoided the magnetic field indicated simply that they were aware of it. What was more interesting was that the animals seemed capable of homing in on certain magnetic positions within their tank until these were artifically altered.

Though the investigators did not feel their efforts were conclusive, it was a step in the right direction. They had uncovered clues that linked these animals' sixth sense with the earth's magnetic force fields in a way we had never previously considered. It meant that these animals might have their own built-in compasses enabling them instinctively to orient and navigate throughout the oceans of the world.

With this new awareness, a host of other questions might be asked. How many other marine creatures are using this unique sense? How about the long-distance homing instincts of certain eels, or salmon, or sea turtles? Are these animals able to tune in on geomagnetic forces that are providing them with necessary information for their phenomenally long-distance navigation and homing feats?

Perhaps. Tomorrow's research should tell us more.

5

How to Meet a Mermaid

Sadie is her name and playing around is her game. As mermaids go she is no raving beauty, but to the average amorous male manatee, she probably represents about two thousand pounds of loveliness. To me, she represents a lovable lady I met swimming one day in Florida. If it wasn't love at first sight, it was a close second. Our affair was brief, but most memorable. One doesn't meet a mermaid every day, but then one doesn't have a love affair with a seacow often either.

Only in the last few decades have men been having any serious affairs with these aquatic mammals. Before that, we simply admired them from a distance. Columbus reported seeing some of Sadie's early family members in 1493, less than three months after he first sighted the New World. His chronicler wrote: "Wednesday the ninth of January . . . when the Admiral (Christopher Columbus) went to the River of Gold, he saw three mermaids . . . they were not as beautiful as they are painted though they have something like the human face. He said he had seen some before off the coast of Guinea . . ."

Columbus's mermaids are believed to have been the West Indian manatee (*Trichechus manatus*), the serene marine mammal also called a seacow. Even after Columbus, certain nearsighted seventeenth-century sailors still thought they were mermaids. But one suspects that, like Columbus, they had been long at sea.

Seacow, a name acquired more recently, links these animals with their very earliest prehistoric ancestors, the elephants. The experts who have studied the geneology of these animals say that one member of the elephant family apparently ran off to sea and never landed after that. They say there are certain skeletal similarities that support this hypothesis. With an order name like Sirenia, from the Latin *Siren*, one wonders if these are the sirens sailors have been harking to ever since Ulysses's time.

I first met Sadie when I was diving near the main spring at Crystal River, Florida. Suddenly, there she was, just sort of hanging in the water like a huge gray barrage balloon, the sun dappling her back in wave shadows. Her size was something to see. It stopped me in my tracks. Though she was alone, I knew others were in the area because it was a cold winter day with freezing air temperatures. But the water temperature, as customary in all Florida springs, was a mild seventy-two degrees. Sadie and her companions were wintering in a balmy environment. As air-breathing mammals easily susceptible to catching colds or worse in the winter, they instinctively knew the spring would keep them healthy until the warmer weather came.

My appearance seemed to startle Sadie as much as hers did me. We both stopped swimming and eyed each other suspiciously, waiting for the other to react. Sadie—and I did not know at the time that this was her name—acted bashful, turning her big frame sideways and watching me out of the corner of a tiny eye.

Abruptly, scurrying around from behind her came a miniature version of Sadie, a young wrinkle-faced seacow gamboling through the water like a playful colt. I suspected it was Sadie's calf. Without the slightest apprehension, the youngster came up to me and playfully nuzzled the coiled cord of my underwater camera strobe.

I reached out and petted her. She in turn arched her back, seemingly enjoying it. I scratched the calf under her chin, and like a puppy, she promptly rolled over on her side to give me more of her to scratch.

At that moment something nudged me from behind. Turning, I saw Sadie at my shoulder. She had pushed her head against me, backed off, and was looking at me out of the corner of her eye as if wondering how I would react to her.

Her nudge was a friendly gesture, but she was cautious about coming any closer. I had a feeling she was not yet quite sure how I would respond. The impression I had was that she was bashful but anxious for attention.

The moment I reached out and touched her, whatever shyness she had before, immediately vanished. As she drew closer so I could pet and rub her almost hairless, rather tough hide, her all-seeing tiny eyes closed tightly and it was no stretch of my imagination to see that she was luxuriating in the gentle massage I was giving her jowls.

Then, almost in slow motion, this great, grayish, round dirigible beside me slowly started rolling over. It was like watching a ponderous iceberg turn bottomside up. The feat finally accomplished, Sadie lay belly up on her back before me, displaying her ample physiography for all and sundry to see.

I rubbed my hand across her chest several times, then caressed her flanks. Her whole body seemed to shudder deliciously. I wondered if I had tickled her.

My hand moved back to her ample bosom. Sadie seemed to stretch out even more, if this were possible. Her head was thrown back so I could tickle the bristly triple chins beneath her bewhiskered lip pads. Then, as I slid back to that broad bosom again, she did something I had never before known a manatee to do. As I rubbed her chest, her two big flippers, which manatees use like hands, reached up and pressed my hand to her chest.

It startled me because it was so unexpected. Did she want me to stop rubbing, or did she like it so much she wanted me to keep my hand where it was? I was a bit helpless because my left hand clutched camera and strobe.

With my free hand clasped tightly to her chest, Sadie slowly started to roll away from me. As she did, I was forced to go with her. It was a weird sensation because she was pulling me up and over the side of her huge body. Since I was using just mask and snorkel, a vision flashed through my mind of how this bizarre turn of events might end up—with me on the bottom beneath this two thousand-pound oddly behaving female seacow. Not my idea of fun.

But she had clasped me so tightly that, when push came to shove, it took all I could do to pull my hand free before she accomplished whatever it was she had in mind. Then she did the ponderous iceberg trick again, rolling rightside up, a mischievous look in her eyes.

Later I learned that it was all in fun, just one of the things Sadie likes to do. Whenever we met—and we had these clandestine meetings several times a year for three years—it was always the

Up to her usual enticement, Sadie rolls onto her back to have her chest rubbed. Next, she may clasp the diver's hand to her breast and execute a slow roll calculated to draw him in, over and under her for a more intimate embrace.

same. First a friendly pat on the head, then a tickle, and Sadie rolled over like a pup. After a brisk belly rub ending at her chest, she pinned my hand, and with (I swear) a grin on her face, she began the slow roll—never toward me always away from me.

When I finally screwed up enough courage to let the lady have her way, it wasn't as bad a trip as I'd thought. I just went up and over, then under, with Sadie settling down on top of me. I reached up and pushed her elephantine body aside and had no difficulty. She was amazingly buoyant. She might weigh a ton out of water, but in it she was about as heavy as a big bag of wet feathers. Her body fat and her large lung capacity probably accounted for it. This explained how the huge mammals bounce in slow motion to the surface every few moments to breathe and make it look like the easiest thing in the world.

Sadie and I became great friends. She was always playful and anxious to please. Often we swam side by side. If I stopped, she stopped. If I held out my hand, she would put her flipper in it. I would then draw her near to take a closeup photograph of her, smiling as only a manatee can smile. Each time the camera shutter clicked, Sadie blinked. She could clearly hear the almost inaudible sound, one apparently not too inaudible to her because, each time it happened, she flinched. Though it did not otherwise seem to annoy her, afterward I would usually put the camera equipment aside and become more involved in developing a relationship with her.

In all our slow, gentle touchings and tumblings together, Sadie's calf stayed close to us, occasionally begging for attention, too. It was almost impossible to devote full time to belly-rubbing the upside-down mother without also having to pacify the little calf jostling beside us. Occasionally, however, the calf would go off to play with the other manatees in the area. Or it would simply settle to the bottom and contentedly graze in the thick pastures of hydrilla weed that grows abundantly on the bottom of Crystal River.

The herbivorous mammals sometimes literally eat themselves into caves of this vegetation, then fall asleep right where they are. Naps are seldom more than five or ten minutes; the mammal surfaces to breathe and then returns to the lush green pastures for another few minutes rest.

Sometimes while swimming with Sadie, I purposely fell behind to watch her. Her large spatular tail had been slashed into three separate pieces by a boat's whirling propeller. The accident may have occurred on an occasion when she surfaced to breathe. Similar marks and cuts appear on the backs of most of the slow-moving mammals. Ironically, these livid white scars, often resembling the serpentine pattern of the trail of a sidewinder rattlesnake across a sandy desert, have helped scientists identify the various manatees.

Usually, when I fell behind and Sadie realized I was missing, she would stop swimming and wait for me to catch up. If I failed to appear, she would then turn around to see where I was. Sadie accomplished this operation by paddling vigorously with one flipper, as might a paddler sculling water, until her great body slowly pivoted around and she could see where I was. If I still remained there and she saw I was not coming, she would then swim back to join me. With one sweep of her broad tail she was quickly beside me, peering quizzically into my mask as if to ask, "What's wrong?"

The seacow was identified and named in the late 1960s by Daniel Hartman, a young biology student who had come to Florida's Crystal River to study the manatees and prepare his doctoral dissertation on them. These mammals are Crystal River's most unusual visitors. They come to this area each winter seeking the warm waters of the head springs in King's Bay. They remain in the general area of the bay with its temperate water conditions until springtime, when they can continue their estuarian meanderings along the coast.

Hartman, at first, was no more familiar with these animals than anyone else. He was only assured from a few local divers that they were peaceful. But even peaceful creatures might be alarmed enough to stampede over a stranger in their midst. So it was with considerable apprehension that Hartman donned mask, fins, snorkel, and dry suit and eased into the spring for the first time near a group of manatees feeding on aquatic weeds.

As Hartman swam cautiously toward the huge, gray shapes, he kept telling himself over and over again that manatees were harmless vegetarians. At that moment, one of the animals looked up from its grazing, saw him coming, and immediately swam toward him. Said Hartman later, "Scientific inquiry be hanged! I swallowed my pride and rocketed back into the boat!"

He was a safe, self-confessed coward, concerned a bit now about how he was going to carry on research if he was afraid of his test animals. Somehow, however, he overcame his fear and before long was developing a relationship with a number of the forty to fifty manatees that each fall returned to Crystal River's warm waters.

Each day, weather permitting, Hartman took a small boat, motored out into the bay, and swam with these ponderous animals for several hours, trying to make himself as unobstrusive an observer as possible. He watched their activities and recorded them on a waterproof plastic chart. On such occasions he was often joined by an enthusiastic teenager named Buddy Powell. Powell had earlier managed to develop a rapport with the manatees and now he offered to help Hartman in his research.

As Hartman and the manatees grew more used to each other, he began thinking of them as individuals. Almost every one of the manatees bore evidence of their encounters with motorboat propellers, the characteristic series of S-slashes showing up as vivid white scars on their backs. By these marks Hartman was able to tell the animals apart. Then he began to name them. Since they were southern animals he gave them southern names, many deriving from the books of William Faulkner and James Agee. Others were named for local people in the area: Creolla, Mae Ella, Jesse Dean, and Cruselle.

He named one female manatee Lavaliere, which is a pendant worn around the neck. (Hartman learned that at the University of Florida students not only got pinned, engaged, and married, but before that they got lavaliered.) Lavaliere turned out to be the most affectionate animal in the whole group, and extraordinarily tame. While most of the mammals were wary of Hartman's approach, she, on the contrary, was extremely friendly. "Lavaliere used to come up and chew all over my dry suit with her thick lip pads," said Hartman. "I never let her take my hand into her mouth as far as her molars. I thought it might hurt. But I used to let her take it in half way. It's an interesting sensation . . . very strong suction . . . you know they have bristles there and this is how they feed. They close the bristles and lip pads on the vegetation and funnel it back into their mouth. There's this constant motion and she would do the same thing with my fist."

Hartman said that Lavaliere was a notorious flirt, brazenly fond of kissing his mask and chewing on his hands and diving suit. On one occasion she managed to undo the front of his dry suit and give him a chilling bath as it flooded with spring water. At other times she embraced him with her flippers and often tried to pull him underwater with her.

Despite the seeming sexual connotations of these maneuvers, Hartman believed that certain of the more friendly animals, such as Lavaliere, were simply expressing their curiosity about the flippered stranger who had come into their environment. For manatees, on a whole, are very gregarious animals given to play among themselves and often with inanimate objects they find in their underwater world. Among themselves, for example, manatees will "kiss" muzzle to muzzle. Hartman also observed that they participate in a large repertoire of nuzzles, nibbles, nudges, butts, and embraces.

Whenever a herd of the animals was engaged in this kind of behavior, it almost appeared to evolve into a serene slow-motion ballet of posturings and positionings, twistings and turnings, with much touching going on at all times. When I first saw this activity six years after Hartman's study, it reminded me of Walt Disney's pneumatic tutu-clad corps de ballet of hippopotomi that nimbly bounced its way through the Dance of the Sugar Plum Fairies in *Fantasia*. The seacows similarly amazed me with their maneuverings, surprisingly delicate for animals of such generous proportions.

Hartman was not the first to notice that the manatees bore individual marks. In his description of congregations of manatees in the Miami River, biologist J. C. Moore wrote in 1946: "One of the first things discovered about the manatees was that some of them were distinctly marked. They displayed prominent scars, great notches on their tails, and distinctive arrangements of lesser scars and notches.

"There were, of course, many individuals that were unscarred. There were also some that could be distinguished for the day by the amount and arrangement of barnacles and brownish marine growth on them . . ."

Hartman drew identification charts showing individual manatee markings, until he could individually identify at least sixty-two of them in this manner.

Communication between Hartman and the manatees was basic, yet it surely existed. He found that one of the animals knew him so well that she always made a beeline for him or his boat. This was, of course, the affectionate Lavaliere who would approach before Hartman even got in the spring. She would nuzzle the propeller and the hull, seemingly in an effort to entice him into the water. Buddy Powell told me of a similar event with a young male named Huggams: "One day I went out and Huggams came up to the boat to where I could reach down and scratch him. After a while I left the area. It was a real cold day and there were about a half dozen boats of fishermen in this area. I left and went around the island and came back five minutes later, stopping the boat in a different place. Within minutes, Huggams was up to my boat again.

"After a while I left again, returning in a half hour to park in another place. Within fifteen minutes he had found me again. This time I reached down and scratched his back with a boat hook which he always seemed to enjoy."

Not long after he completed his manatee study and wrote his dissertation, I interviewed Hartman for a magazine article. I asked him what manatee talk sounds like. At that time I had never been in the water with any of the animals.

"It's chirps, squeaks, squeals, or screams—sounds apparently emitted more in an emotional context than in an intercommunicative context," he said. "The animals vocalize in situations of alarm, distress, male sexual arousal, and less to alert a group. It is questionable whether the manatees are evolving towards greater sociability or away from it. I think they are evolving away from it, so there is not this bond, this vocalizing bond, let's say, that dolphins have where they exchange a lot of communication of social significance, of social value. I've seen the animals squeal . . . calves squeal while the other animals are around and they just don't care. They are not attracted by the calves squealing in fright. The one predictable social duet is that between a cow and its calf. Every time you see that, it's beautiful. It always happens that the calf will get scared and squeal and the mother will immediately answer. Or the mother will squeal and the calf will immediately answer its mother. So you get this beep, beep . . . beep, beep . . . until finally they come over and investigate the disturbance and then they split."

I asked Hartman if this could be a homing type of call for each other. He said, "Yes, I'm sure that one of the two partners, usually the calf, homes in on where the mother is calling from and goes right to her side. Once it is there and the cow knows it is there, she is apt to stop squealing. But when she starts to move out, to swim away, to cruise, or to flee, then she starts squealing again because then the chances of losing the calf are increased. Also, I have the suspicion based on several times when I dove in turbid waters, that vocalization is increased under these conditions, again, to assure contact so they won't lose each other . . ."

"When they are playful," I asked, "did you see them making any type of communication or sound to indicate that they are enjoying themselves?"

"Yes," said Hartman. "But I cannot differentiate between the sounds. They all sound the same to the human ear. I'm sure if we get these on a sonagraph sometime we can probably find some good variations."

"Do you think they use the sounds in any sense of ranging?"

"No. I don't think they echolocate. I don't think they use their vocalizations for navigation."

Since this was in the early 1970s, Hartman felt that the growing number of divers who visited the Crystal River Springs area was a deterrent to the manatees. To be able to study the animals at leisure, he himself had to race against the first divers that would show up in the spring every morning. Sometimes they tried to ride the manatees and rather than let the animals approach, they pursued them. Consequently, the manatees were driven from the spring area out into the chillier waters of the bay. Now this might not seem so critical, but the air-breathing mammals can catch pneumonia this way and die.

With the increased numbers of divers who have come to Crystal River Springs through the years, the manatees have received more attention than usual. When Hartman first worked with them, only a few were friendly enough to develop a close relationship with him. But since then, with more divers constantly trying to pet or caress the animals, those mammals wishing a relationship with the divers establish it quickly. Apparently, this is happening in fairly large numbers, more so today than ever before. A part of the wintering herd here have become strongly sociable with the growing clan of divers and it has created a problem: the animals are in danger of being smothered by affection. When you see two or three animals surrounded by divers all of whom are trying to touch and caress the seacow at the same time, the scene is reminiscent of a teenage idol being overwhelmed by admirers.

In the years after Hartman left, I often went diving in Crystal River with the manatees. I found that the Crystal River dive-shop operator Charles D. Tally knew a lot about the animals. Tally had worked with Hartman on some of his research projects and he had made some interesting observations. Tally said that he could tell the manatees apart not only by their individual markings but also by their different personality traits. For example, he said, "Piety is a young female seacow. If you get in the water near her she will come racing to you just as frisky as can be, but she is very bashful. She'll come up and tuck her nose in the sand and hide from you. She'll want you to look her straight in the eye, but she just can't hold a steady gaze with you. She gives in. When you turn and start to go away from her, she'll reach out, grab your flipper and hold you back. If you are climbing into the boat, she'll be right there at your flippers, begging you to come back in. You can always recognize Piety by her identification marks—just forward of her tail on the right side, she has six white propeller scars like a row of S's. Even if you couldn't see her identification marks you could identify Piety by her personality.

"Lockinvar is a big male that Woody [Hartman] got chummy with. Lockinvar is the only one that ever became any kind of a threat to Hartman, if that's possible. They used to roll and tumble together, playing. And one day Lockinvar grabbed Hartman by his shoulder-length hair and pulled him to the bottom.

"Woody thought, 'Well this is the end of that. This ol' boy is getting too rough.' He said that just about the time he was running out of breath, but not yet in a strain to breathe, Lockinvar turned him loose.

"I've experienced that before with these manatees," said Tally. "They seem to know just how long a man can stay underwater.

"Sewer Sam is the one Jacques Cousteau brought in. He has a horizontal *K* crease on his nose. The base of the *K* is close to his eyes with the *V* pointing toward his nose. He'll come to you whenever he sees you and he'll also come if you start bouncing an oar on the bottom of the boat. He'll hang around until you get in the water with him."

The curiosity of the manatees for unfamiliar things in their underwater world is best exemplified by an experience Tally described involving a diver who came to the springs one day to film the seacows.

"Since he was a double amputee," said Tally, "he used an underwater propulsion unit [a battery operated torpedo-shaped unit with handles and propeller used to tow a diver underwater] to get around. He was out there filming the manatee with this unit and the animals were fascinated by the propeller sound of the scooter *underwater* with him dragging behind it. This was something that most of them had never seen before, so they were curious. This fellow was playing with one manatee in particular when something happened to his camera case.

"He was wearing scuba, so he set his scooter down on the bottom and was about six feet away from it, checking his camera equipment when this manatee he had been playing with came up to him, looked him squarely in the face, then swam over, stood looking over her shoulder at him, then with her nose she went from one end of the scooter to the other, nuzzling it.

"Then, while he watched, the manatee picked the entire scooter up in her flippers and hugged it to her breast as if to say, 'While you're busy, I'll play with your baby.' "

As anyone who has spent time in the water with manatees knows, the mammals are very adroit with their flippers. Not only do they use them independently to help them move about in the water, but they help bring food to the animal's mouth and in this sense are used as crude hands. You can feel fingerlike bones within the flipper. They possess nails that look much like elephant toenails.

Often we see manatees pushing their large lip pads and cheeks with their flippers, apparently in an effort to dislodge a piece of vegetation caught between their teeth. Most of the animals are too apprehensive about allowing a diver to place his hand near their mouth or face; however, one young seacow I met one day relished this kind of contact.

It was initiated by her coming up and kissing me full on my face mask. In the process, she placed both flippers on each side of my head to accomplish it. I, in turn, reached out and put my hands on each side of her head and gently stroked her cheeks. I was rubbing her whiskers and had worked down around her thick lip pads when she started nibbling on my fingers. Not sure if I was ready yet to be gummed by a nibble-happy calf, I held up my hand flat in front of her face and she continued nibbling my palm. Her prehensile lips reached out, curled over my fingertips and held my hand firmly in place where she could nibble better.

This particular calf enjoyed all manner of petting and was fond of being touched and rubbed, especially around her flippers and under her jowls. Sometimes she would head for the bottom in a series of spirals, then come back to the surface to see where I was. Unlike some manatees, she enjoyed making close frontal contact. One area that seemed in particular to please her was under the apex of her flippers. If there is any sexual connotation to the man/manatee relationship, part of it may stem from this particular caress, for the apex of the flipper is where the nipple is located, one on each side, and where the female suckles its young. The nipples are capable of erection, which occurs when a diver touches them. Not all female manatees solicit this caress, but most do, and the contact is quite obviously pleasurable for them.

Like myself, Charles Tally is a strong believer in the similarity between manatee behavior and that of man's best friend. "He is just like a puppy dog," said Tally. "He comes up to you and he's bashful, wagging his tail, wanting to play with you, but he doesn't know whether he can trust you until you reach out to pet him. He really doesn't know whether you are going to hit him, step on him or play with him. But once you let him know that you are going to pet him, he'll eat you up. That's the way these manatees are. Once the contact is made, they have no qualms about rolling over on their back and letting you rub and pet them. The only time they will come up to breathe is when they run out of air and have to breathe. Some will trust you enough and just stick their noses out and breathe, while others will swim away to breathe, then come back to you. This is one reason why you should never follow them; let them come back to you."

As friendly and playful as the manatees were during the day, I wondered how they would react to a diver at night. One New Year's Eve, a friend and I decided to find out. We loaded scuba gear into our boat and headed for the springs to see if we could find the animals and have a nighttime relationship.

It was a cold, drizzly New Year's Eve. About 10:30 P.M. we quietly paddled our boat to the far end of the spring run, an area known as the Grand Canyon, where manatees often congregate during the day, basking in the flow of the main spring. Careful not to cause an undue disturbance, we donned our gear and slipped into the water. The beams of our underwater lights quickly revealed that the Grand Canyon was empty; there were no manatees.

Skirting this white-sand area of the main run, we swam underwater toward the springs, staying just above the thick beds of bottom vegetation, wondering if the animals might have bedded down in the hydrilla for the night. Halfway to the main spring we apparently frightened a mother and her youngster, because mud suddenly exploded before us and I glimpsed a pair of spatulatic tails—one large and one small—beating swiftly through the water as the animals made a fast getaway. We made no effort to pursue them, knowing full well that the only way to relate to these animals is to be patient and let them come to you.

These were the only two manatees we saw that night, although we continued on into the spring, inspected the basin, and swam into the rock cavern, about thirty feet deep, from which the main source originates. The fact that we saw none of the animals where they usually spend their daylight hours suggests that they may at night frequent areas not normally visited by divers. Night dives in the spring are popular with scuba enthusiasts and it would seem logical that the manatees would prefer sleeping where they were less likely to be disturbed.

Hartman's keen observations of the manatees' behavior resulted in a large amount of provocative data concerning these unique animals. His swimming with them, playing with them, and recording the events in their lives every day for over a year has provided us with some idea of the day-to-day activities of these peaceful mammals and their interspecies relationships.

"More than once," said Hartman, "I observed what appeared to be play-soliciting behavior. A juvenile cow seemed to be luring bottom-resting companions to play as she rolled on her back and grazed their sides. Similarly, a young bull swimming on his back made several lazy approaches to bottom-resting cows. Another young male chewed on the tip of a cow's tail while she bottom-rested, but elicited no response.

"Occasionally, manatees play by themselves. An adult bull broke from the playful embraces of two juvenile males, slid along the bottom on his belly, then rolled on his back, skimming the sand and plunging through vegetation. Calves amuse themselves by twisting, tumbling, and barrel-rolling through the water. While his mother was resting twenty meters away, Huggams, completed several minutes of play by 'rocketing' to the surface so that his chest and flippers broke water."

Biologists are uncertain as to the function of manatee play. Moore suspected that the often-seen act of manatees touching each other muzzle to muzzle in the typical manatee "kiss" attitude was a behavioral carryover from their terrestrial ancestors, the elephants. He felt that it was probably a form of greeting, perhaps a method of mutual identification or recognition. From Hartman's observations, however, when two manatees met they did not engage in any kind of formal identification behavior. In fact, they hardly glanced at each other, and if two were approaching head-on, they simply veered aside to avoid collision. Hartman felt that the play between animals in which nibbles and nuzzlings were involved was a kind of grooming often seen in primates, and that it served simply to strengthen the basic, loose social bond between members of a herd. Still, Hartman postulated that there was more behind the mouthing between these animals than met the eye.

Though it was sheer conjecture, he wondered if manatees possessed sense receptors in their tongues and responded to taste cues. "Do manatees, for example, possess a chemical sense or a 'smell taste' by which they can recognize odor gradients in the water?" he asked. "Can they obtain directional information from salinity gradients?"

Was it possible, he wondered, for male manatees to tell when a female was fertile or infertile by the taste of their hides? Or was it a water-borne scent that communicated this information? Was there an "identifier" factor that enabled individuals to distinguish one from another?

Hartman knew that biologists David and Melba Caldwell believed that sperm whales actually laid a trail of scent in the water which provided chemical information for other whales and assisted them to maintain contact between members of a pod. Does a comparable communication system exist among Sirenians, he wondered.

Hartman was impressed by the apparently strong bond between mother manatees and their calves. According to Moore, "Immediately after delivery the mother manatee lifts her calf above the surface of the water on her back and then dunks it repeatedly until it establishes a breathing rhythm of its own." Within half a day of their birth, Moore observed calves capable of swimming and surfacing by themselves without their mother's assistance.

During the first few days of its life, a new-born calf of a captive seacow was seen to swim entirely by its flippers and occasionally hitch rides on its mother's back. From one to two years, the parent and its offspring will maintain a strong bond, the calf constantly reinforcing this relationship by mouthing, touching, and kissing its mother. The two maintain close contact at all times except when a herd of seacows are together; then the calf may wander off to play with other manatees.

At such times, one of the things that Hartman was curious about was whether the cows and calves ever separated beyond their range of effective communication. During his observations he noted that one of the six- to eight-month-old calves belonging to a seacow identified as Flora Merry Lee seemed more independent than most calves and frequently strayed far afield. Yet, the cow and calf always managed to locate each other easily, sometimes from as far away as 180 feet, at which time the calf, Burgeon, returned quickly to the mother when alerted by her squeals.

Hartman felt sure that manatees could communicate underwater over even greater distances. He noted that the only kind of defense that a cow and her calf resorted to when menaced was to scream an alarm and flee.

"A cow answers scream for scream any defense calls from its calf, hurries to it and leads it away from the source of anxiety," said Hartman. "Miss Molly, for example, called her calf to her side as it was about to surface too near a boat. A bull rising from the bottom collided accidentally with Bess's calf, who screamed in surprise, igniting a duet with its mother and bringing her to investigate. One calf called Tenbrooks swam squealing to its mother when pecked by a sheepshead." This species of fish is commonly seen following the manatees, especially right behind their tails or hovering over their backs. They have sharp teeth designed for feeding on vegetation and crustaceans on pier pilings. But for the Crystal River sheepshead in this brackish water, an additional source of food is found in the algae growing on some of the manatees' tails and backs. Often in its eagerness to snap up a morsel, however, it irritates the manatees into full flight.

Watching the manatees relate to divers in the spring, Hartman noticed that, even in cases where the young were frightened, the animals never displayed any agonistic behavior. Once, however, he did see a cow apparently dissuade her calf from investigating a diver: the mother, Flora Merry Lee, simply nudged her calf aside gently as it was approaching Hartman.

During the courtship period, calves try to remain close to their mothers, but are often pushed aside by the more aggressive bulls. During this intense activity, when the mother must frequently flee from seven or eight amorous males, the calf can get left behind. Hartman saw the mother double back to find the straggler on numerous such occasions.

"A cow persistently pestered by bulls sometimes flees without summoning her calf to her side," he said. "Deserted, the calf either waits until called by its mother or swims screaming in the direction of her departure."

One might get the impression that these large lumps of animals, seemingly slow-moving and benevolent, are cowlike in nature and not very intelligent. In part this is true; yet the sensory capabilities of the mammals are quite remarkable. Hartman noted that the manatees possess exceptional acoustic sensitivity. Sound is undoubtedly the main factor determining many of their social activities. By vocal cues alone, the bulls are able to head straight for the herd from a considerable distance away under extremely turbid water conditions. They are also adept in localizing surface noises. Even when the water is so turbid that they cannot see the splashing of divers, they quickly home in on the source.

By splashing water on shore with his hand, Hartman attracted animals from over 45 feet away. One, passing about 120 feet from his boat, was seen to alter course to investigate Hartman bumping an oar against his boat's gunwale. From the air, others were seen to alter courses and zero in on the main herd from over 150 feet away. Hartman also found out how sensitive the animals were to manmade noises. He noted the same thing I had observed, that the click of a camera shutter a few meters from the head caused the manatees to wince. "Once," Hartman said, "a bull lolling beside me flinched when I blew water out of my snorkel; and often the animals were bothered by the divers' noisy exhalations from their scuba gear." This is one reason why the animals are less frightened and more apt to approach divers when they are using only mask, snorkel, and fins.

Hartman saw idling manatees wince in unison from the sound of an outboard motor changing gears thirty feet away. Another time a cow flinched at the splash of a diver entering the water ninety feet from her. Even sounds outside, beyond the surface, bothered them: the roar of a low-flying jet fighter caused a resting male manatee to flinch violently on the surface and dive for the bottom.

A manatee's eyesight is apparently fairly good, for we have had animals recognize us at distances of over one hundred feet in clear water and seen them turn and come our way. Like some other marine animals, manatees have good low-light vision, the eyes shining pink in the dark from reflected light indicating this capability, although Hartman said the animals rarely shied away from a flashlight beam, contrary to our one observation on the night dive.

The manatees' broad backs are covered with short, bristly hairs that stand upright like tiny antennae. When we petted the animals on their backs, we saw ripples run across the entire area as if it were tickling them. Hartman hypothesized that these dorsal hairs might be receptive to low-frequency vibrations and pressure waves generated by the swimming motions of these aquatic vertebrates.

No research has yet been done to determine if certain sounds attract and others repel the mammals. Tally felt sure that, as with dolphins, some manatees are attracted to the whirling sound of boat propellers. Hartman, following manatees in his runabout as they swam along the river, said that on at least one occasion the animals had paid so little attention to the sound of his outboard motor that when one of the creatures surfaced to breathe, it collided with the bow of his boat. Quite possibly the mammals grow so used to the sound they hear so often in this area that they are no longer afraid of it.

Like many marine mammals, however, different sounds in their environment may cause more than passing curiosity among the animals. For example, Tally told me that sometimes he sat on the bottom wearing scuba gear and humming a tune in his mouthpiece. In a few moments, he said, several manatees would gather around him, settle to the bottom, and seem to be listening to the tune with the rapt attention of a well-behaved audience.

Ten years after my initial interview with Hartman, I located and interviewed Buddy Powell, Hartman's young assistant who had gone on to Stanton University to major in wildlife management. He had returned to Crystal River as a member of the National Fish and Wildlife Service to continue working with the manatees in that area. Some of the manatees that he remembered from his early days were still visiting Kings Bay at Crystal River. One of these was the female named Piety. Powell had been keeping track of this seacow's offspring for at least two generations. I asked him if any of Hartman's hypotheses about the manatees had not held up. He told me that Hartman had not believed the manatees were as social as they seemed to be. But Powell's daily observations and seacow counts by airplane seemed to indicate they were far more social than Hartman had thought. Rather than being loners, they were often in the company of at least one other member of the herd.

The manatees exhibited the ability to follow each other through the same underwater corridors when the animals were perhaps a mile apart, even when the water was murky. Powell said: "We still don't know how they do it. They seem to be operating on a sixth sense sometimes. I've watched manatees swimming in the St. Johns River when it was so muddy there was hardly six inches of visibility. I was able to follow one by his surface swirls and I saw that animal change course and swim diagonally across the river to zero in on a random floating hyacinth, which he ate. Now how do you explain that ability? I've seen the animals do this repeatedly and it made no difference whether their targets were upriver or downriver; they could make a beeline for them without even seeing what they were aiming for, and they were able to do this over a distance of a couple hundred yards."

When I asked Powell if the upright hairs on the manatees' backs might serve as some kind of sensors, he said he thought they probably did. As early as the 1800s it was noticed that the dorsal hair follicles were evenly spaced a fair distance apart and that the nerves in this dorsal area suggested that the hairs were tactile and capable of picking up some kind of perceptible messages.

In order to learn just how sensitive the backs of these animals might be, Powell stuck one with a pin. The animal never flinched, as he might have from a small clicking sound such as a camera shutter. There was no visible reaction whatsoever. This suggests that the manatee's sense of hearing may be more acute than its sense of feeling, at least on the skin.

In recent years, Powell has been attaching thumb-sized transmitters and so-called spaghetti tags to the backs of manatees to study their migratory habits. After a few hours, the animals forget the sonic tags are even there. From monitoring these transmissions, Powell has learned that, when the manatees leave the springs and begin their normal wanderings up and down our southern coasts, they may travel as many as thirty miles overnight; their journeys carry them as far north as the Virginia–North Carolina borders.

No matter how far they range, caring people like Buddy Powell will not be far behind in their efforts to study and preserve this vanishing species. Manatees are endangered animals and need all the protection they can get if they are to survive as a species and not be slaughtered into extinction as was their cousin the Stellar seacow.

The last winter I visited the seacows I had my most memorable experience. That morning my wiper blades froze to the windshield, the car was coated with ice, and at nine A.M. as I slid into the water near the spring, the air was so cold the water smoked.

It was the day after Christmas and northern Florida was weathering a cold snap. That night the mercury had nose-dived into the twenties. It was just what I wanted; the colder the better. I figured the manatees would be bunched up in the spring and it looked like a good day for relating with them.

Mask and snorkel adjusted, camera and strobe ready, I finned away from my boat and watched the brown rocks and hydrilla slide past below my mask. First the hydrilla disappeared, leaving only the rugged bare rocks, then they too vanished, leaving only the deep green void of the spring.

From its sun-splintered depths, a school of jacks drifted up in a slow-motion swirl of silver shapes, riding an upcurrent from the spring cave that fractured the cliff face thirty feet below.

I paused to watch their unhurried ballet before moving on across the deep pool. Just as I reached the buttressing boulders on the far side, I saw my first swimming companion of the day: a half-grown manatee wedging herself beneath the broad shoulders of a bottom fissure, rubbing her flanks lazily against the rocks and her stomach on the soft sand bottom. At first I failed to understand what she was doing. Then I saw that she had placed herself precisely in the middle of the spring run and was luxuriating in the warm water flowing past her from the spring pool.

As I drifted down the run, looking into the distance, I saw several manatees emerging from the green-amber murk like great gray barrage balloons. Three appeared on my left, all adults. Then two calves and a female emerged to my right. Several more shapes materialized behind them. It looked like quite a herd.

When they saw me they swam over to investigate. They encircled me, large and small alike, cheek by jowl with each other, their tiny button eyes winking as they looked me over. The calves were more curious and less cautious than their elders, as usual, moving in close as if to sniff me out.

"Hello!" I burbled through my snorkel, holding out my hand to them. "Hello, you marvelous mermaids!"

They responded by moving closer. Before I knew it, calves, cows, and bulls were crowding me from all directions. I felt like a traffic cop caught in a busy intersection with a broken traffic light.

The herd apparently accepted my presence without a second thought, for at one point I was wedged tightly between three animals and found myself from necessity studying at close range the evenly spaced sensorlike hairs standing straight up on a pair of broad backs. Enough algae grew in patches further back to attract a pair of hungry sheepshead. In their eagerness, one fish nipped a cow's back harder than usual and instantly—one apparently frightening the other—both manatees bucked furiously away from the finny molesters, much to my relief.

With the traffic jam cleared, I turned to find a friendly calf hanging at my elbow as if waiting patiently for some attention. Slowly, I reached out and petted his wrinkled hide. Here was another animal with this bad skin condition that makes them look older than their years. The hide was creased and wrinkled as if with age. Some of the calves are often so wrinkled from one end to the other that they resemble woolly sheep.

After photographing the youngster nibbling my fingertips, I moved back among the elders and began snapping pictures, recording a variety of back scars I had never seen before. Moving in among the mammals, I noticed one in particular that kept following me. Finally, I turned to snap her picture, then was shocked to see she had but one flipper. The other had been neatly severed just ahead of the teat. How, I wondered. A boat propeller, nylon fish line, shark, alligator? All were there in the river, even the large sharks that venture into this fresh and brackish waterway from the Gulf of Mexico.

The rest of the manatee's body was surprisingly free of any other marks. To assist in my examination of her, she obligingly rolled belly up. I patted her chest. She reached up with her one good flipper and pressed my hand to her bosom. It was a simple, touching gesture.

After a while, my left hand holding camera and strobe began to cramp. When I tried to withdraw, she clamped down tighter. Shades of Sadie I thought, thankful that at least she wasn't trying to roll me. Tugging only drew her to me. Somehow we ended up in a

SMILE! would seem the appropriate caption as my one-flippered manatee friend appears to strike a pose for the underwater cameraman. This Crystal River, Florida, manatee was especially fond of relating to divers, seeming anxious to share their attention and affections.

kind of cheek-to-cheek embrace. And she still had my hand.

Curious to see just how long she would tolerate this familiarity, I started swimming her around on her back. She seemed charmed. Her head back, eyes closed, a slight "smile" on her lips, she struck me as looking terribly anthropomorphically content with herself. Together we waltzed around the spring run together. It would have made quite a picture. The rest of the herd acted as if we were something they saw every day. Either that or they were too blasé to look.

After a while, I was the one to break the embrace. My cooperative partner rolled upright and seemed satisfied to lead me back toward the main herd. I followed beside her. As long as I stayed there where she could feel my presence, she kept swimming. Once, however, when I stopped to take a photograph of her from behind and she got ahead, I saw her stop, look right, then left, trying to spot me. Then, vigorously, her one good flipper began pulling at the water. Slowly she turned around until her little eyes caught sight of me. Then, like an overgrown puppy, she bounded back to my side. Only later, after I had slipped back across the main spring to my boat to change film, did I momentarily lose sight of her.

Returning to the spring run again, I encountered my old friend Sadie and her calf. It was a nice reunion. While I was playing with her calf, Sadie watched me soulfully. Finally, she nudged my shoulder for a little affection, too. Reaching out, I petted her head. That's all it took and my friendly dirigible turned bottomside up. But the ploy didn't work this time. I had had enough hand-holding. When she saw I would not play her game, Sadie rolled rightside up, eyeing me as mischievously as ever. With a friendly pat on the head and a final tickle of her chins, I waved good-bye to Sadie and finned off to see what was happening with the others.

The herd grazed contentedly in the thick pastures of aquatic weeds growing beside the sandy run. I watched and photographed them for several hours, occasionally taking time to play with the friendly calves or with my one-flippered friend.

At about two P.M. I witnessed a sight few people have ever been privileged to see before. Three manatees I had not seen earlier swam into the run. A large female was followed by a young male. Tagging between them came a calf belonging to the female. The persistent male kept nuzzling the female. She in turn seemed willing to accept his advances. Before long, she pushed her muzzle into the soft sand bottom, lifted her tail toward the surface at a forty-five-degree angle and the male, approaching from behind, wrapped himself about her for the conjugal embrace. The two remained in this position for seven minutes, then separated, both surfacing briefly for a breath of air. Immediately afterward they resumed their embrace and this activity, periodically interrupted for breathing, continued for the next hour.

Meanwhile, the female's calf remained nearby, never straying more than a few feet away from its mother. Sometimes it sidled up

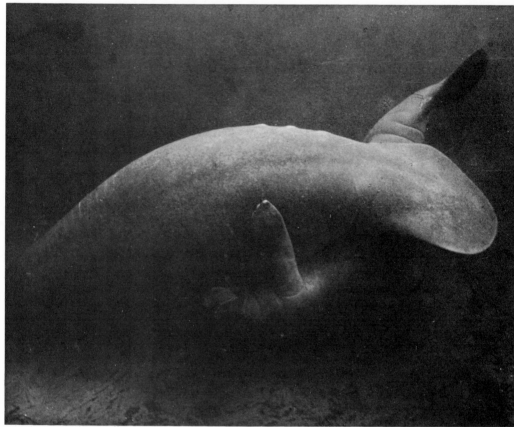

Seldom seen or ever photographed, the conjugal embrace of these two manatees is the ultimate expression of communication between these mammals. While they coupled, the female's calf drew alongside and began suckling at the mother's side.

beside the mating pair as if seeking attention. After nuzzling the mother's flanks several times, the baby got what it wanted. In the act of mating, the mother shifted her flipper and the offspring eagerly pushed its muzzle over the exposed teat and commenced nursing.

For me, this scene summed it all up. Here was everything man had come to love about these marvelous mammals. Here was caring and sharing, playfulness and peace. Here was Sirenia, the mermaids of mythology, their silent siren song coming through loud and clear.

6

The Talkative Killers

In the mid-1950s, off the rocky coast near Southern California's Palos Verdes cliffs, scuba diver Bob Marx, bored with spearfishing, headed for the bottom to photograph some bright red Garibaldi fish. Nearby, several sea lions played beside some rocks. As Marx swam toward them, hoping to take their pictures, the animals suddenly screamed and began acting frantic, some trying to climb out of the water and onto the rocks. Marx thought his presence had frightened them, then suddenly he saw the real cause for alarm—hurtling out of the green gloom beyond the rocks came four killer whales, each about twenty feet long!

Instantly, Marx realized the predicament he was in. Killer whales were believed to be among the deadliest animals in the world, more cunning and powerful than the largest man-eating sharks, and here he was treading water between the killers and their prey!

"I just froze," Marx said, "sinking like a stone to the sea floor about thirty feet down, and this was probably my salvation. Luckily I had plenty of air left in my tank, for as I looked up, the killer whales closed in, expertly cutting off their victims' line of retreat to the shore. Seconds later the sea was a churning caldron of blood. Smaller sea lions were cut in half with one crunch of those massive jaws; larger ones, some weighing up to five hundred pounds, were finished off almost as quickly with three or four bites. In two minutes there was nothing left of the twenty-odd sea lions but gory bits and pieces that filtered down on top of me. Fearing that the killer whales might notice the tidbits they had missed, and me in the bargain, I quickly set out for shore, hugging the bottom like a snake. I was so

127

rattled by the horrifying scene I had just witnessed that I left my camera and its case, which had taken me a month to build, on the bottom and was never able to locate it again."

It has long been believed that no man has ever survived a killer-whale attack and lived to tell about it. Yet, it seems that the infamous reputation of this so-called killer, this marine mammal known scientifically as *Orcinus orca,* is due largely to no more than two or three accounts that stress the villainous characteristics normally associated with typical "bad characters"—viciousness, violence, cunning, and the rapacious appetites of a cold-blooded killer.

Early twentieth-century Antarctica explorer Robert Falcon Scott seems to have single-handedly struck the mold that would shape all killer-whale terror tales to come. Since Scott's single diary entry, made during his fatal last expedition to Antartica in 1911, has so totally influenced our concept of killer whales right up to and including the present time, it is worth presenting here in its entirety.

> 1911, Thursday, January 5—All hands were up at 5:00 this morning and at work at 6:00. Words cannot express the splendid way in which everyone works and gradually the work gets organized. I was a little late on the scene this morning, and thereby witnessed a most extraordinary scene. Some 6 or 7 killer whales, old and young, were skirting the fast floe edge ahead of the ship; they seemed excited and dived rapidly, almost touching the floe. As we watched, they suddenly appeared astern, raising their snouts out of the water. I had heard weird stories of these beasts, but had never associated serious danger with them. Close to the water's edge lay the wire stern rope of the anchor, and our two Esquimaux dogs were tethered to this. I did not think of connecting the movements of the whales with this fact, and seeing them so close I shouted to Ponting who was standing abreast of the ship. He seized his camera and ran toward the floe edge to get a close picture of the beasts, which had momentarily disappeared. The next moment the whole floe under him and the dogs heaved up and split into fragments. One could hear the single "booming" noise as the whales rose under the ice and struck it with their backs. Whale after whale rose under the ice, setting it rocking

fiercely; luckily Ponting kept his feet and was able to fly to security. By an extraordinary chance also, the splits had been made around and between the dogs, so that neither of them fell into the water. Thus it was clear that the whales shared our astonishment, for one after another their huge hideous heads shot vertically into the air through the cracks which they had made. As they reared them to a height of 6 or 8 feet it was possible to see their tawny head markings, their small glistening eyes, and their terrible array of teeth—by far the largest and terrifying in the world. There cannot be a doubt that they looked up to see what had happened to Ponting and the dogs.

The latter were horribly frightened and strained to their chain, whining. The head of one killer must certainly have been within five feet of one of the dogs.

After this, whether they thought the game insignificant or whether they missed Ponting is uncertain, but the terrifying creatures passed on to other hunting grounds, and we were able to rescue the dogs, and what was even more important, our petrol—five or six tons of which were sitting on a piece of ice which was not split from the main mass.

Of course, we have known well that killer whales continually skirt the edge of the floes and that they would undoubtedly snap up anyone who is unfortunate enough to fall into the water; but the fact that they could display such deliberate cunning, that they were able to break ice of such thickness (at least 2½ feet) and that they could act in unison, were a revelation to us. It is clear that they are endowed with singular intelligence, and in the future we shall treat their intelligence with every respect.

Typical of the kind of information that has resulted from Scott's description and which still persists in our literature today are these statements found in a contemporary book describing dangerous marine animals:

> The killer whale . . . has a reputation of being a ruthless and ferocious beast. . . . Killer whales hunt in packs of 3-40 individuals, preying on other warm-blooded marine animals. They are

fast swimmers and will attack anything that swims. They have been known to come up under ice floes and to knock seals and people into the water. If killer whales are spotted, the diver should get out of the water immediately. . .

Paul Spong, a New Zealand psychologist who has been studying whale behavior since the late 1960s, spent years observing and working with wild orcas in Vancouver, British Columbia. He said that, in all the time he had studied these so-called "killers," he knew of only two cases in which orcas had actually attacked and killed people. One occurred in 1956 when two loggers who were skidding logs down a slope into the water saw a group or pod of killer whales passing nearby. According to the account, one of the loggers intentionally dispatched a log, which slid down the slope and struck one of the whales. Though the animal was not killed, it was apparently injured. The whales left the area.

That night the two loggers were rowing back to their camp when suddenly the family of killer whales appeared beside them. They bumped the boat and tipped it over. One of the loggers never came to surface. The other, untouched, lived to tell the tale. The missing logger was the one that had loosed the log that struck the whale.

Was this accident intentional or coincidental? Did the logger simply drown or was he gobbled up by the retaliating orcas? From such stories legends are born. And in their retelling, the truth is sometimes lost.

Today, fortunately, man is beginning to come out of the dark ages of his understanding about these marine mammals, thanks largely to individuals who have dared to confront the "killers" on their own terms in their own environment and lived to tell about it. Indeed, they are rapidly dispelling our ignorance and fear of the unknown and replacing them with significant truths about *Orcinus orca*, the largest member of the dolphin family.

This whale of a dolphin is best characterized by its rounded head, statuesque dorsal fin, large conical teeth, and unique black and white markings. With the exception of a white eye patch, the upper parts of the cetacean are glossy black, the under parts stark white. This unique pattern of colors suggests that Mother Nature had a

Some scientists believe orcas have vivid, high-contrast coloration for these reasons: flank white spots may enable pod members to see each other more easily; head markings may be disruptive coloration making the head difficult to distinguish for the orca's prey; the white belly markings seem to point out the genitals, or in the females, the nipples. The killer whales in bottom illustration are standing upright with heads out of water to survey their surroundings, a maneuver called "spyhopping." After *Mind in the Waters*, assembled by Joan McIntyre, Charles Scribner's Sons, New York, 1974.

definite reason for their arrangement. Largest members of the clan grow to about thirty feet long. The male's erect dorsal fin often exceeds six feet in height, a slender black blade so tall that its uppermost tip may lay over almost at a right angle. Since these dorsals and the manner in which their tips turn over sideways vary considerably, they serve as a kind of identification mark for whale watchers, enabling them to identify individual animals.

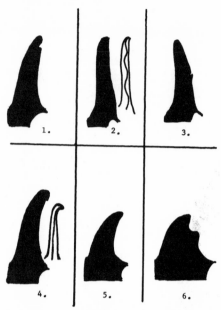

The peculiar shapes of some orcas' dorsal fins help identify various animals for whale-watchers. These silhouettes show the variety. All are drawn as though the whales were facing to the left. (1) A female adult named "Nicola" for the notch near the top of her dorsal fin. (2) A senior bull called "Wavy" for the waves in the leading edge of its dorsal. (3) "Hooker," or "Forward Fin," is the name of this adult male because of the characteristic forward slant of the dorsal. (4) "Floppy" is the obvious name for this whale whose tall dorsal tip flops over. (5) The shortness but complete fin shape here marks a baby whose dorsal fin is not yet full-grown, as opposed to (6) an adult female orca whose fin may have been lopped off by a ship's propeller. Its abbreviated form earned its owner the name "Stubb." After *Mind in the Waters*, assembled by Joan McIntyre, Charles Scribner's Sons, New York, 1974.

Living together in family groups called pods, orcas are believed to remain with their families throughout their lives. No one has determined positively the life span of these animals, but authorities believe they may live to be at least eighty years old.

Their diet consists mainly of fish, sea lions, seals, and sometimes porpoises, sharks, and squids. Their choice of food probably depends upon what is most commonly available in their area. Orcas range throughout the oceans of the world and in all seas are active animals. Occasionally they attack large whales, forcing open the mouth of these great creatures to get at what observers believe must be a delicacy high on the orcas' menu: the unfortunate animal's tongue.

In 1978, aboard the research vessel *Sea World*, belonging to the Hubbs-Sea World Research Institute of San Diego, Captain Robert Vial and others witnessed a vicious attack on a sixty-foot-long blue whale by a pack of about thirty killer whales. As later described by *National Geographic* staff member Cliff Tarpy, the attack had already been in progress at about one P.M. and it appeared that the killers were trying to strip away flesh and blubber piece by piece from the fleeing leviathan that was trailing a river of blood.

"Some might think killer whales, tame and playful in captivity, unfairly named," said Tarpy, "but in their habitat, killer whales do kill. The predators exhibit distinct divisions of labor; some flanked the blue on either side, as if herding it. Two others went ahead and two stayed behind to foil any escape attempts. One group seemed intent on keeping the blue underwater to hinder its breathing, another phalanx swam beneath its belly to make sure it did not dive out of reach. The big whale's dorsal fin had been chewed off and its tail flukes shredded, impairing its movements. The dominant bulls led forays to pull off huge chunks of flesh.

"The attack continued until early evening. *Sea World*, covering nearly 20 miles, followed the struggle for five hours. But its total duration is unknown since the spectacle was well underway when discovered. . .

"About 6 P.M. the attack came to a halt—suddenly and mysteriously. First the killers toward the rear slowed down. Then those toward the front turned back, and they all swam away."

Why? Perhaps their appetites were sated, or perhaps they were waiting for the great blue to weaken even further. No one really knows; but in this case the killers left the scene and the probably mortally wounded blue whale continued swimming, with strips of bloodied white blubber where its dorsal fin had been, long ragged wounds where skin had been stripped from its torso, and one gaping cavity over six feet wide that had been gouged in its side.

As with most predators, orcas are fast and clever. Large animals capable of preying upon any kind of sea life, they themselves are not preyed upon by any creature other than man. The estimated mortality rate of these animals from commercial fishermen and others often acting in ignorance and fear is several hundred a year.

Orcinus orca inhabits the cold coastal waters from British Columbia to California with such frequency that their migrating pods are common sights along the West Coast during certain times of year. They appear less frequently elsewhere, but seem to visit most of the world's oceans. Wherever they show up, there is no doubting their presence. A glimpse of their glossy black rounded backs topped by the upright six-foot dorsal blades has, until recent times, struck fear in men's hearts. But in July 1965, an event occurred that did much to change worldwide public opinion about the reputations of these killers. A twenty-two-foot-long, four-and-a-half-ton, male killer whale was captured accidentally by commercial fishermen off the coast of British Columbia near the town of Namu.

What followed became an American legend. Essentially, it is the story of one man's effort to save a whale and to try and develop a relationship with it. It is the true story of amateur naturalist Ted Griffin, who appealed to the people of Seattle, Washington, to help him come up with the sum that was being asked by the fishermen for the whale. Practically overnight, the local citizens responded to Griffin's plea and the deal was made.

Griffin paid eight thousand dollars for the whale, which he named Namu; then he built a floating pen for the young male killer and towed Namu 450 miles to a wharfside oceanarium in Seattle. The public welcomed the whale with the warmest show of affection any populace has ever afforded a killer whale.

Griffin began to study the large mammal. Day after day he spent hours on the catwalk of the whale's floating cage, watching it, noting its behavior, observing its moods. Then, gradually, he began to approach Namu. At first he entered the pen and rowed a small boat about inside the enclosure. Displeased, the animal went to the bottom and sulked. Gradually, however, it began to accept the man and his boat.

Griffin switched to an inflatable raft and continued coaxing the whale to approach him. Soon, Namu was allowing Griffin to stroke and pet him. Before long, he was tagging along behind Griffin and his rubber boat like a faithful dog. At this point, Griffin decided to risk his life on his own hunches and get in the water with the killer.

Wearing a black wetsuit against the chill he slipped into one corner of the pen and ducked his head underwater to see how the whale would react, and also to keep the big animal in sight in case he suddenly seemed aggressive and Griffin had to exit hurriedly. But everything appeared to be all right.

"I could see the burly black-and-white form holding position with a lazy rocking motion fore and aft," said Griffin. "His eye held mine. Gradually I felt secure.

"With a short-handled brush, I approached Namu underwater. With a light touch I scrubbed his head, nose, and chin. The whale made no move to withdraw or attack me.

"Later the same day I swam right around Namu. He stayed stock still, seeming almost to disregard me.

"I slid onto his back and tugged gently at his dorsal fin. He swam two or three times around the pen with his unfamiliar burden. . . . From that point on it was like a honeymoon. We got along beautifully."

The bond between man and whale grew rapidly. Namu so enjoyed being groomed by Griffin, having his whole huge body gone over briskly with a short-handled scrub brush, that it would grow almost ecstatic over the experience. The two played games together inside the pen. One favorite was a variation of piggyback in which Namu would roll upside down, reach up with its disc-shaped pectoral fins, clasp Griffin's boat to its broad chest and swim around the pen on its back with Griffin riding in the boat. The crowds on the pier loved it.

Namu never tired of this kind of play. But occasionally Griffin did. Sometimes the only way he could bring the game to a halt was to throw a salmon far enough ahead of the boat to entice Namu to release it, and while the whale was enjoying the tidbit, Griffin rowed like mad for the catwalk.

At other times, Namu would allow Griffin to swim upon his back, holding onto the tall blade of the dorsal fin while the whale romped around its enclosure, the two creating a spectacle that no one had ever seen before.

Everyone who had worked with Namu up until then was well aware that the animal was vocalizing both above and below the water. His beeps and squeals intrigued Griffin, who had noted earlier that during the long float trip down the coast Namu had been in communication with other killer whales. The animals were obviously signaling back and forth to each other. Although other whales had approached his comparatively flimsy floating pen, no effort was made by them to rescue Namu. Yet, a female whale and a youngster repeatedly swam in concert with the penned Namu all the way down the coast. Griffin theorized that these were possibly Namu's mother and younger brother or sister. All three animals were in constant communication throughout the entire trip.

Before the journey actually began, before the two became such close friends, Griffin had tried imitating the same sounds when he was with Namu, more in an effort to pacify the big animal than for any other reason.

"At Warrior Cove," said Griffin, "I entered the water close to Namu, but still with the net between us. The whale peered curiously at me and here took place our first eyeball-to-eyeball talks. . . . I tried for the first time to match his sounds. This proved easier underwater, for some reason, than in the air. By the time we had Namu in his pen he was responding to me. Admittedly, these were strange conversations; I hadn't the faintest idea what I was saying to him with my own feeble vocalizing."

As time went on, Griffin began to realize that he was dealing with an extremely intelligent animal, one that showed a considerable measure of craftiness, quite characteristic of this entire family of toothed whales, as indicated by one distinctive incident. Dolphin handlers can almost duplicate this kind of anecdote, but in Namu's case it was a first.

It began one morning when Griffin and his assistant noticed that Namu seemed hungrier than usual. He kept accepting each salmon that was passed out to him, and repeatedly came back for more. His appetite seemed so insatiable that Griffin suspected the whale was up to something. To find out exactly what, he donned his wetsuit and slipped into the water.

"It took only moments to discover a pile of fish stacked like cordwood in a corner of the pen," said Griffin. "Namu was hoarding rations, apparently for a quiet salmon break between regular feedings!"

Thousands of people came to see the man and whale and the unique relationship that had developed between the two. Among them was Thomas C. Poulter of the Stanford Research Institute. He was interested in making underwater recordings of Namu's vocalizations. He told Griffin that Namu created sound underwater with his larynx, but when he was vocalizing on the surface the sounds came from his blowhole. What fascinated Griffin most was that, although he had heard Namu vocalizing repeatedly, listening to Poulter's tapes of these sounds, it was impossible for him to detect any duplication of sounds or duplications of groups of sounds.

Once Namu's chatter was analyzed by the scientists with electronic sound-analysis equipment, they learned that the whale was capable of transmitting sounds over a wide range from 50 cycles per second to at least 40,000, perhaps even 100,000 cycles per second, while humans are limited to a range of 50 to 20,000 cycles per second. This meant that they were able to hear only a portion of the sounds that Namu was emitting. The question that came to mind for these researchers was whether the whale was actually using the information he got from generating such high-frequency sound waves.

Griffin and the others were soon to marvel at Namu's incredibly selective sonar, which the whale was able to demonstrate for the scientists when Griffin tried to get Namu to accept an albacore tuna rather than his favorite foodfish, the more expensive salmon. The only way they were able to lure the big whale into taking the tuna was to dip the salmon into the water until the whale rose up for it, and then substitute the tuna.

Griffin continued this experiment at night when he knew Namu could not see the fish. He held the tuna in the water about twenty-five feet away from Namu and dunked it up and down. Namu paid no attention; but as soon as Griffin dropped a salmon in the water, the whale swung his broad head back and forth, apparently scanning the target with his sonar, then slowly swam over and accepted the choicer offering.

"He was making a lot of audible sounds and probably more that we couldn't hear," said Griffin. "The two fish were about the same size and shape. To me it was a striking display of the killer whale's ability to discriminate and select solely by the use of sonar."

Gradually, as their relationship became closer, it also developed another element. Griffin became aware that, despite the benign nature of the whale up to now, Namu had moods that indicated he might not always be the tame, docile "pet" that Griffin seemed to have made of him.

"The whale knows more about me, too," said Griffin, "for instance, that I am slow in the water and speak his language atrociously. Today my friend, tomorrow he could pick up habits of disdain, carelessness, or even aggressiveness that could imperil our relationship.

"Though I try not to let the whale know it, I've discovered that he has a mean side. I have taught myself to beware when he swings his head sharply or jerks it upward. Especially, take care when he bobs his head up and down swiftly. This marks his most severe displeasure and peevishness. 'Leave me alone now,' it says, and I take the hint."

Griffin's experiences with Namu changed the whole character of our concept of these animals. As one noted cetacean authority, A. Remington Kellogg, observed, "I'm afraid we must toss away some of our earlier preconceptions about these animals. This behavior of Namu is entirely contrary to what anyone could have expected. I would be cautious, though, about generalizing from the actions of this one killer whale. I wouldn't use my trust in this animal as a passport to familiarity with another."

Dr. John C. Lilly, famed today for his dolphin research, suggested in numerous conversations with talented underwater film maker Ivan Tors after 1962 that orcas, despite their unsavory reputations, would probably accept a relationship with man in the water with them. Dr. Lilly felt sure that the large animals would react to man in the same way the smaller species, dolphins, did, and that they would be accepted.

Later, Ivan Tors made a movie about Namu and spent six weeks filming the whale that Griffin had brought to Seattle. As we all know, the results were the charmingly delightful Walt Disney film *Namu, the Killer Whale,* which introduced us for the first time to the gentle, compassionate nature of what might well be one of the most maligned creatures in the sea.

Disney's film did more to change our thinking about these animals than any single event. Everyone, young and old alike, fell instantly in love with the squeaky-voiced hero in the oversized penguin suit, and from that moment on, the trend has continued. Today, killer whales are favorite performers at some of our popular oceanariums, and to all who have had the opportunity of catching their acts, it is apparent that they show the most remarkable unkillerlike behavior one could imagine. But as Kellogg pointed out earlier, it is not wise to take this animal for granted.

As man began to make friends with this reputed "savage killer," he gradually became more interested in the sounds the animals made underwater. What did it all mean? Was it possible that these large marine creatures could be trained to perform the most complicated feats? Did they actually have sufficient intelligence to communicate vocally with each other? Were these squeaks, squawks, creakings, and groans some kind of "whale language"?

Not only from Namu, but from killers at large as well, the scientists began to record the sounds these animals made. The tapes were analyzed and it was soon discovered that, rather than random noise, these sounds were clearly signals made for some special purpose. Some signals were thought to be used for echolocation to help find their way in unfamiliar areas; other signals were believed to be communicative in nature, used for the sole purpose of exchanging information between animals.

The signals were analyzed on an instrument that provided researchers with a picture of these sounds, a sonagram, that was as revealing to the investigators as a fingerprint would be to the FBI. Here were actual pictures of the killers' language!

When another killer whale—a young female named Shamu—was captured, researchers were twice as excited by the possibilities. When scientists introduced Shamu into the same cove with Namu, it was an excellent opportunity to record the sounds exchanged by two captive killer whales. Researchers placed three hydrophones underwater with the whales and commenced recording on three channels of a tape recorder. A fourth channel was employed for a running commentary by observers of what was occurring between the animals at the time that they were vocalizing, their relative positions to each other, and which animal was sounding off at that particular time.

From these tapes, the scientists learned that neither the male nor the female duplicated signals. Analysis of the picture record of the sounds revealed that specific characteristics of the signals indicated whether the male or female was making the sound. This meant that, by the sound alone, a listener could tell whether it was a male or a female vocalizing. Interestingly, however, as the investigators were analyzing the signals, a faint ghostly echo appeared on the tape. It was made by another female killer whale; yet, no other animal was in the cove with Shamu and Namu!

When special triangulation listening devices were used in an effort to locate the source of the faint overlapping signals, researchers were astonished to find that the signals came from beyond the cove, from an area named Rich Passage. Apparently, a wild orca was sounding off out there and the signals were being picked up not only by Shamu and Namu, but also by the underwater hydrophones. This explained to the researchers why the two whales often vocalized at times when they were not involved with each other. Apparently, they were commenting or communicating with this distant voice!

The mystery vocalizations came from three miles away. Since these were being blocked by a nearby point, the scientists felt sure that killer whales could probably communicate over distances greater than seven miles. With this knowledge, they went back and examined earlier sound tapes in which Namu seemed to be vocalizing by himself. Now, for the first time, they noticed the same sound in the background, the faint response of a wild killer whale. Normally, solitary orcas were not known to vocalize. It was nice to know that Namu was not just talking to himself.

Over a thousand killer whale sonagrams underwent computer analysis to record and isolate individual signals. Between midnight and morning on two consectuve days, Namu emitted 210 signals, 189 of which were analyzed by sonagram; the remaining 21 were eliminated because they were obscured by background noise.

After all the analyzing, recording, and isolating of signals, all the experts could say with any certainty was that the animals possessed an extremely complex system of signals easily recognizable against any background noise. It was a means of communication which the orcas could modify by accenting, abbreviating, punctuating, syllabifying, hyphenating, prefixing, and adding other numerous endings and inflections—all without destroying the signal's initial character so that it was still easily recognized. But why the animals modify signals in this manner is still a mystery to man. This is only another way of saying that so far we have been unable to break the code.

"On the other hand," said investigator Poulter, "I suspect that the different signals do make sense to other killer whales. If we were to consider the sonagram charts to be sonagrams of Namu's vocabulary and we had the nearest equivalent that exists in the English language for each of these signals, I suspect that we would still be orders of magnitude away from being able to make any combination of them that would make sense, either to us or to the killer whale. . . . The system of signals of the killer whale seems to supply the strongest argument (from the standpoint of communication theory) that 'animal language' is statistically valid; we feel that killer-whale signals are more important in this regard than the signals of any other species. We believe that marine mammals talk and what they talk about makes sense to other marine mammals of the same species."

Playback experiments of other scientists brought about the following results: Paul Spong and his companions visited Alert Bay, British Columbia, in January 1970 to observe and record the presence of orcas in the Johnstone Straits region of British Columbia. While there, they performed various experiments to determine whether orcas in their natural habitat would show any interest in a variety of taped sounds. Placing loudspeakers above and below the water, the investigators bombarded the local whale population with every kind of music and pure tones on every frequency, various incidental sound patterns, playback recordings of vocalizing orcas and other whales, and delayed feedback of the orcas' own sounds.

"All to little avail!" wrote cetologist Spong. "The orcas consistently displayed (to our minds) remarkable indifference to our efforts to catch and hold their attention. Although they showed some interest occasionally, they mostly seemed to ignore the sound stimuli we projected at them. The most interesting reactions occurred in relation to feedback of their own vocalizations. When the feedback was simultaneous, they virtually stopped vocalizing. However, with delayed air feedback, the whales produced some very unusual vocalizations, seeming to play with the procedure. In general, we concluded that the whales were usually preoccupied with the routine hunting activities when they passed by our station, and that the types of sound stimuli that we were using held little interest for free whales."

If free killer whales were unresponsive to their own and other orcas' vocalizations, this was not the case when other whale species were treated to orca "talk."

In 1969, during the annual whale migration off San Diego between mid-December and mid-February, an interesting experiment took place. While an estimated 11,000 gray whales were moving southward toward warmer water breeding grounds, a single member of that group was seen traveling alone at a speed of about five miles an hour. Suddenly, the underwater stillness was shattered by the screams of a pack of attacking killer whales.

Instantly, the gray whale turned and plunged into a thick forest of kelp. To all appearances, the ponderous mammal had vanished. Not even its "blow," the stream of ejected warm air that appears over a whale's head when it exhales on the surface, was seen. Apparently the animal had exhaled underwater, for in a few moments the whale rose cautiously to the surface in the kelp to quietly inhale and look around.

Where were the screaming killers? They never existed. The whale saw nothing nearby more threatening than a sailboat. A Navy catamaran lay alongside the kelp. It was the source of the killer whale screams. Aboard the sailboat Navy scientists had broadcasted the sounds underwater to observe their effects on the lone gray whale. They believed the animal took refuge in the kelp forest because the kelp's gaseous air bladders may impede or distort sonar signals killer whales use while hunting.

The scientists repeated the experiment with other whales. Some dove into the kelp while others simply reversed their course and swam northward. If no killer whale sounds were broadcast, the gray whales continued their placid southward journey. This was also true when the scientists broadcast a variety of random sounds and noises underwater. The whales paid no attention.

"One of the intriguing things about this playback experiment," said author Flora Davis in her book *Eloquent Animals, A Study in Animal Communication,* "is that it amounted to communication from one species to another, using the signaling system of a third species. To date, scientists have no idea what killer whale screams actually signify, and probably gray whales haven't either: they may be simply recognizable as the voice of *Orcinus orca,* the killer whale."

More recently, killer whale feedback has been used for a far better purpose than to test its response on migrating gray whales. In 1978, in the normal course of gathering their catch in the waters of the world, Japanese commercial fishermen killed over a thousand dolphins. More than just the mortality rate of the dolphins was in-volved. Japan had a considerable problem to cope with. The coun-try's food supply mainly consists of fish and other marine creatures. In their efforts to farm the sea, the Japanese fishermen found that great numbers of dolphins were attracted to the same schools of fish attracting the fishermen. If the dolphins did not simply eat much of the fish, they chased them away or damaged Japanese fish-catching devices. The Japanese estimated that this deprivation of their fish-eries was costing them between three million and four million dollars a year. So it seemed imperative that they come up with a way to prevent the wholesale slaughter or dispersal of their fish, while at the same time restraining their natural impulse to eliminate the predaceous dolphins that were simply doing what comes naturally.

But how? How would you control the natural appetites of great numbers of hungry dolphins determined to steal the fish that you are after? When one Japanese fisheries expert recalled that dolphins were sometimes on the killer whales' menu, it followed that the sounds of killer whales might somehow drive off the marauding dolphins.

So incensed had the dolphin lovers of the world become over the killings in the Japanese fisheries that the Japanese government hired dozens of experts between 1978 and 1979 and invested $185,000 in a unique effort to solve the problem. For years the fishermen had tried to capture the dolphins and move them from the area. This, however, had only resulted in higher numbers of fatalities. Though the dolphins were often clever enough to avoid live capture, they frequently succumbed to harsher methods. One bit of knowledge seemed a step in the right direction. The Japanese knew that dolphins sent out sound waves for echolocation. They also suspected that they were communicating among themselves by emitting other sound waves. Through experiments with dolphins in captivity, the researchers learned that striking two iron pipes to-gether underwater effectively drove off dolphins. Coupling this information with the fact that schools of dolphins were known to try to vacate an area immediately when they detected the sounds of orcas, even if they were far away, the scientists tried to learn how they might most effectively use these sounds to drive off dolphins without doing the same to the schools of foodfish they were after.

Eventually, the Tolkei District Fishery Laboratory and a Fishery

Devices Research Team recorded a variety of dolphin signals and the effects certain sound waves had on the dolphins and the other marine creatures of the fishery. From that information they developed an economical noise generator that emitted a combination of sounds that repelled dolphins but did not affect the commercial fish.

Then, while Japanese sound technicians completed a forty-six-minute continuous cassette tape containing these sounds, another group built a lifelike model of a killer whale that was to carry this equipment "to the foe"—the predatory dolphins.

This lifelike orca replica was in itself quite an achievement, consisting of a fiberglass-reinforced body 13 feet long and 7½ feet high and weighing 396 pounds. The model orca was then primed with its vital, waterproofed, electronic components—a 24-volt d.c. battery, a 31-pound tape recorder, a 44-pound noise generator, a 57-pound loudspeaker, a 31-pound, 12-volt, battery power supply, fore and aft ballast tanks—and the monster was ready for its sea trials.

On the continuous loop tape was the sound of killer whales made in 1964 at Vancouver, British Columbia, and from a captured orca whose sounds were taped in 1978 in a Japanese marina. The "false killer" proved adequately seaworthy to depths of 150 feet. To avoid any engine noise that might upset the dolphins, the orca replica was to be towed.

Meanwhile, the team involved with taping the sounds for the continuous loop tape hit upon the best way to prevent the orca's signals from panicking the commercial fish. When researchers found that dolphins communicate at certain ultrasonic frequencies at least one hundred times higher than the lower frequencies that might bother fish, they had the key to the problem.

The next step was to see how captive dolphins reacted to the various sounds created by the scientists. The tests took place at a large aquarium where dolphins were trained and fed. First, the scientists struck the iron pipe while the dolphins fed. Immediately, the feeding ceased. It was apparent that the animals were highly annoyed by this noise, so much so that they refused to feed. Meanwhile, the scientists recorded the exchange of signals between animals at this time.

Next, fifteen dolphins were exposed to four different types of noises and their reactions were observed, videotaped, and re-corded. First, the metallic noise in the 1,000 to 2,000 Hz range was tried; next came ultrasonic sound waves in the 2,000 to 4,000 Hz range; and third, the test animals listened to the playback of orca sounds and dolphin "speech" responses during that time. Finally, the large monster orca, with all of its noisemaking apparatus blaring, was brought into the area with the fifteen dolphins.

One can imagine the kind of response this generated in the test animals. Observers reported that the dolphins reacted most vio-lently to the orca signals and the ultrasonic sound waves. Those nearest the orca and its blaring speakers raced to the opposite side of the fiberglass whale, toward shore, where they could be furthest from the sound. The animals that were spread over a wide area immediately clustered together. Some that had been swimming in the depths surfaced.

While the research was not conclusive, observers felt that the response in the dolphins was as had been anticipated—precisely what they were trying to achieve. However, other related experi-ments were intended, and with an annual expenditure of about $190,000, the project was expected to be refined and soon in use. Its sponsors hoped it would send out sound messages that would repel dolphins from the fisheries. A projected spinoff from this idea may be the next development—corraling large schools of fish with spe-cial noise and sound-wave ultrasonic-signal "fences" designed to frighten off predatory schools of dolphins, while at the same time keeping the great masses of commercially valuable fish in check. Only one small detail still bothers the scientists. How long will it take, they wonder, for the highly intelligent dolphins to learn that they are being duped and fail to respond as anticipated?

Only after the public was caught up in the popularity of Namu in the mid-1960s did anyone consider the possibility of working with and perhaps training the intelligent killer whales. Eventually, such West Coast oceanariums as Marineland of the Pacific and Sea World near San Diego, California, accomplished this feat. Today, audi-ences are literally spellbound by the leaping leviathans that "tail-walk," do headstands, wave at the audiences with their flukes, carry their trainers astride their backs around circular tanks, or perform the old Clyde Beatty head-in-the-lion's-mouth routine with their brave handlers. At San Diego's Sea World, a trainer rides the back of a killer whale as it dives to the bottom of its deep pool and propels itself upward into the air, leaping clear of the water with the man on its back, diving immediately back to the bottom of the pool again, and repeating this five or six times before finally swimming to the side of the pool so that the trainer can step off safely.

"This is an incredible performance," said Dr. John Lilly. "I could hardly believe it the first time I saw it. . . . This is an astounding cooperative effort on both the part of the human and the killer whale. This man has immense courage and immense trust in this huge creature. On the other hand, the killer whale has an immense trust in the humans and does everything he can to be sure that that man can breathe and does not drown. This requires a discrimination and a careful timing of the dives and the leaps in such a way that the man can survive."

Dr. Lilly felt that, without the excellent organization in such oceanariums as this, these feats would be impossible. "The immense sensitivity of these animals' skin allows them to detect the presence of a person and to regulate their activities in such a way as not to damage them." He added, "It is most impressive, their control of their immense size so as not to endanger their human friends."

After watching a similar program at Marineland of the Pacific, George Reiger, describing the species for an *Audubon* article, told of staying behind in the stands while the other spectators left to catch the sea lion and dolphin shows on the other side of the Marineland complex. With everyone gone, the two killer whales were left by themselves, still standing on their heads waving good-bye to the crowd:

> For a long while they continue the fluke-waving routine. You sit alone in the bleachers and watch them, the huge animals balanc-ing on their heads and swinging their tails back and forth. Some-times the flukes come crashing down on the surface and the particular animal is off lobtailing around the tank, smashing the surface with each rise and fall of its big flukes. Eerily, the splash-ing stops and one or both of the killers rise up out of the water and turn slowly, looking. They appear to see you and pause a moment before sliding beneath the surface and repeating the tail-waving

or lobtailing sequence. Whether they think you are a potential source of food, or whether they are just so devoted to the act that as long as one spectator remains the show must go on, is hard to say. They probably continue their antics impulsively to work off energy and emotion.

Interpretation is everything in our view of nature, and nowhere does it play a more significant role than in our understanding of *Orcinus orca* the largest of dolphins.

As with all other animals in these various oceanariums, trainers are not apt to reveal all the tricks of their techniques of communicating and training their charges, be they orcas, dolphins, or seals. But you can be sure that the procedures used are painstakingly slow at first and involve progressively more difficult and repetitious activities. Hand signals play an important part, as do underwater sounds trainers employ to condition the animals for certain responses, all of which are performed for food rewards; in some cases, simply a pat on the head or an indication that the handler is pleased is enough.

Scientists have found that, as is true with all of the members of the toothed-whale family including the dolphins, orcas apparently see in the air as well as they do underwater, and are capable of instantly correcting their vision for one element or the other. This feat, which humans are incapable of, is a matter of interest to ophthalmologists presently engaged in studying the phenomenon in dolphins.

In addition to their ocular capabilities, the animals possess unique "sonar sight" that enables them to echolocate under conditions where underwater visibility may be nonexistent. Both their own and other sonar signals are picked by the orcas through their rostrum, the brow of their head, and the tip of their lower jaw. Whenever an orca is listening underwater, it orients itself head first to the sound source, much the way we might do in turning our heads slightly aside to pick up sound better with our ears. We now know, too, that these animals are capable of picking up considerable information through sensors on their tongues. This largely has to do with taste and smell. Dr. Lilly describes captured dolphins in their pool in the Virgin Islands positioning themselves in the ocean inflow to their pool with their mouths slightly open and tongues out, sensing not only distant food sources, but probably body essences of their own kind swimming offshore in the area.

Discussing cetacean senses and communication in Sam Houston Ridgeway's *Mammals of the Sea, Biology and Medicine,* investigators David and Melba Caldwell wrote, "We have observed considerable tongue movement by a killer whale, *Orcinus orca,* and have seen it frequently in *Tursiops truncatus* (bottlenose dolphin). We have seen animals of these two species stick out their tongues for no other reason as they swam around the tank. . . . Much as some humans do when solving a problem, young *Tursiops* have sometimes been observed to stick part of their tongue out of the side of their mouth when learning a new routine."

At San Diego's Sea World, Caldwell photographed a young female orca extending its tongue and pressing it against the glass of an observation window in response to the grating sound made by an observer holding one end of a pocket comb to the window and running his fingernail across the comb's teeth. The animal was apparently attracted by the sound and could see the comb.

No one yet knows for sure whether the movable tongues of the toothed whales function as touch and taste receptors, but scientists agree that, from all observations, this certainly seems likely.

For some time, the U.S. Navy has realized the benefits of training marine mammals to perform useful jobs. If we could learn more about how they cope with their underwater world, perhaps this information would help man as he becomes more involved in this watery environment.

With these goals in mind, supported by the Office of Naval Research, a modest facility for studying and working with marine mammals was established at Point Mugu, California. The Marine Bioscience Facility developed a Marine Mammal Training Program involving dolphins, orcas, and sea lions. Probably its most popular and best-known student was "Tuffy," a nickname for Tuf Guy, a dolphin acquired in May 1964 from a Santa Monica seaside amusement park. Almost overnight the lovable Tuffy attained fame and instant popularity as the deep-diving tool and message-carrying member of the aquanaut team led by former astronaut Scott Carpenter during their historical stay in Sea-Lab II, the underwater habitat established in two hundred feet of water off La Jolla, California.

Lesser-known members of the Marine Mammal Training Pro-

During the U.S. Navy research with marine mammals, Ishmael, the killer whale, is lowered toward the floating platform designed to carry him to a floating pen in Mugu Lagoon at the naval research station at Point Mugu, California. Official U.S. Navy photograph.

gram were a pair of killer whales named Ahab and Ishmael, which had been captured off Seattle, Washington, in 1968 and air-lifted to Point Mugu. As director of the Bioscience Facility, Forrest G. Wood described the training program in his book *Marine Mammals and Man: The Navy's Porpoises and Sea Lions*. Both whales were to be used in a Navy project called Deep Ops, designed to demonstrate how these animals could be used to recover various pieces of naval hardware, such as torpedos.

The two orcas received basic training like the dolphins at Point Mugu lagoon. They were on their best behavior and so friendly with their trainers that Ahab would allow his trainer, Jon Hall, to walk about on his back while Hall scrubbed him with a long-handled brush. The killer slowly rolled over in the water to expose unbrushed portions of his seventeen-foot length.

Ishmael often followed his trainer, Blair Irvine, out into the open lagoon, swimming behind Blair's small boat. Both whales had been conditioned to respond to a recall signal.

In 1969 and 1970, the whales were transported separately to Hawaii, where they were joined by a twelve-foot-long pilot whale named Morgan that had been flown there a year earlier. The whales were all fitted with a special kind of head gear featuring a clamp.

"When pushed against a target," wrote Wood, "the clamp was triggered and at the same time detached from the mouthpiece, which the whale then returned to his trainers." Later, the device triggered a gas-inflated balloon, which floated the object to the surface. The whole idea of these tests was to determine how deep the whales could dive while holding their breath and carrying out their assigned task—the recovery of some type of ordinance with the head clamp or grabber.

Interestingly, Morgan, the pilot whale, proved to be a more agreeable deep diver than the killer whales. Although Ishmael would, for example, make a dive to five hundred feet, on a subsequent test he made only a shallow dive, surfaced, spat out the mouthpiece, and slapped the water with his flukes and flippers in apparent anger or frustration. A short while later, the orca swam off and was never seen again.

On another test, the other killer whale, Ahab, refused to make a dive for a target recovery when the target was switched from 500 to 750 feet deep. Moreover, this animal also seemed to have a mind of its own and appeared reluctant to return to its base, preferring instead to wander off on a twenty-four-hour jaunt that covered fifty nautical miles, accompanied by its frantic chase boat. Often, it appeared that Ahab and his chasers were not always going in the same direction. Eventually, however, the wayward whale was rounded up with the assistance of the radio transmitter the animal carried which signaled him back, and finally he docilely returned to his home port, none the worse for the experience.

Since Ishmael had already gone AWOL, however, and Ahab now

showed similar inclinations, the Navy "washed them out" as potential partners in the deep-water recovery program. They directed more attention to their star pupil, the receptive Morgan, who obligingly dived to a record depth of 1,654 feet with the recovery device, holding his breath for over 12.5 minutes. But that was that; Morgan stoutly refused to dive any deeper, with one exception, when the target was at two thousand feet. Unfortunately, however, his depth-of-dive transmitter was malfunctioning at the time and there was no way to determine exactly how deep he had dived. But observers noted that he was below the surface for over thirteen minutes and probably made it all the way to the bottom.

As Wood observed, "Much remains to be learned about the behavior of killer whales, which seem to differ in interesting but ill-defined ways from some other species that we have worked with."

This observation appears to be as valid today as it was then. Relatively little is yet known about the largest of the dolphins, but what little we have learned indicates that the highly intelligent *Orcinus orca* possesses the potential for a meaningful relationship with man when the proper time comes. Every new bit of knowledge, each new research effort, adds new pieces to the puzzle.

For example, in 1980, while studying various pods of killer whales and recording their calls at sea, University of British Columbia researcher John K. B. Ford reported that groups of whales containing up to forty individuals were using a kind of signature whistle peculiar only to their pod. Instead of calling them signature whistles, however, Ford called them stereotype calls, or S-calls. He recorded the S-calls from twelve different pods of whales and found that each pod had a repertoire of about twelve to fifteen different ones. In this area of study, although several pods of whales often traveled together, the individual pods still used their own S-calls. Ford believes that these S-calls permit pod members to maintain contact with each other while traveling with killer whales of other pods. They help preserve a group identity. In eight years of observation, Ford has found that, once a whale has been born to a pod, it never leaves that family group; thus this may be the pod members' method of maintaining contact.

While the researchers would like to learn whether or not the whales are communicating more information than just identification

Opposite: At the Naval Undersea Research and Development Center in Hawaii, Morgan, a pilot whale, answers a sonic cue that will send him on a unique mission for the Navy—the recovery of naval ordinance from the ocean's depths. *Below:* View of the grabber device the whale uses to assist in recovery of torpedoes from depths of one thousand feet as part of the Navy's Deep Operation Recovery System. *P. 154, top:* The twelve-hundred-pound pilot whale approached trainer to receive the grabber device it has been trained to hold by its mouthpiece during the dive. *P. 154, bottom:* Morgan approaches the torpedo and prepares to attach the grabber. *P. 155, top:* After pressing the claw against the torpedo and locking the lift device, Morgan pulls away with the detached mouthpiece. Simultaneously, a hydrazine gas system inflates a balloon that will float the torpedo to the surface for recovery. *P. 155, bottom:* Mission accomplished, Morgan returns to trainer and receives a fish snack reward. Official U.S. Navy photographs.

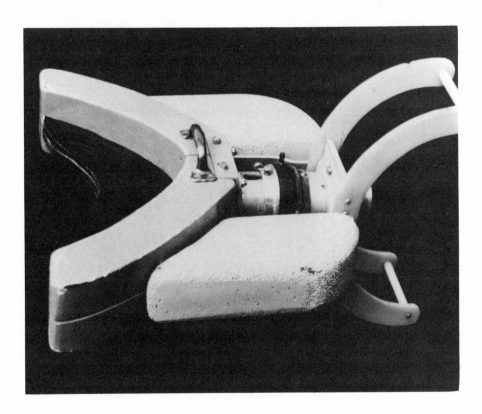

signals with each other, so far they have only found that one S-call has definitely been associated with particular behavior. In this instance, when the group is resting and making slow dives together, they always emit this particular signal.

Future research by Ford and others includes using playbacks of S-calls to see if they cause any particular kind of behavior in the whales. Quite possibly they will, for aerial observations have shown that, while pod members may be out of sight of one another while foraging, they all change direction in unison. Playback experiments and analysis of resulting behavior may one day provide us with the clues to find out more precisely what these killers are talking about.

7

Listening to Leviathans

At dusk, thirty-five miles off the coast of Bermuda, a small sailboat reached across the gentle ocean swells. It was the quiet time of day. Too late now to risk finding their way back through the treacherous reefs, Roger Payne and his wife, Katy, decided to spend the night at sea. Out there on the featureless, gray, sliding facets of ocean, with daylight slowly disappearing, the two were suddenly aware of how alone they were. When you can see to the very edges of your world, and you are the only living creature in sight, your aloneness becomes quite palpable.

As the darkness closed in upon them, the Paynes were sharply aware of the feeling of solitude common to most open ocean sailors. "I felt as one with the other solitary watchers elsewhere on earth," said Payne, "the shepherds, sentinels, herdsmen who huddled alone beneath the same stars, feeling the night close in around them."

But the Paynes were not sailing across the far reaches of the Bermuda waters simply for the fun of it. They were there for another purpose. And now, as Payne swung the boat's tiller and brought the vessel to a virtual standstill, hovering almost directly into the wind, the couple lowered a pair of underwater hydrophones into the sea. Then they put on headphones, switched on amplifiers, and listened.

Suddenly, the Paynes were no longer alone!

"Instead, we were surrounded by a vast and joyous chorus of sounds that poured up out of the sea and overflowed its rim," said Payne. "The spaces and vaults of the ocean, like a festive palace hall, reverberated and thundered with the cries of whales—sounds that boomed, echoed, swelled, and vanished as they wove together like strands in some entangled web of glorious sound. I felt instantly at ease, all sense of desolation brushed aside by the sheer ebullience of it all. All that night we were borne along by those lovely, dancing, yodeling cries, sailing on a sea of unearthly music."

Unearthly is indeed the proper word to describe it. The Paynes were listening to and recording the plaintive song of the humpback whale. This was not a song in the sense of a popular melody that might make the Hit Parade. Rather, it was a song such as a bird might sing, a characteristic, repetitious type of vocalization capable perhaps of many variations. For well over a decade, marine zoologist and cetacean authority Roger Payne and his wife have observed whales at close range and systematically made underwater recordings of their vocalizations throughout the world. As early as 1971, Payne was trying to obtain a record of the sounds made by the southern right whale *(Eubalaena australis)*, found congregating in the western South Atlantic off Argentina's Patagonia coast. Sponsored by the New York Zoological and the National Geographic Societies, Payne, his wife, and photographer-divers Bill Curtsinger and Chuck Nicklin spent three months there studying the behavior of these remarkable whales. The heavy-bodied leviathans attain a weight up to forty tons and a length of fifty-eight feet. It is a slow-swimming baleen whale that apparently enjoys breaching—hurling its huge body into the air with such violence that it sommersaults backward, smashing back into the sea with a thunderous explosion, an activity a right whale may continue twenty or more times, always landing on its back or side.

Part of Payne's research involved getting into the water with the herd and filming the animals at close range. Another phase involved monitoring their underwater sounds through three orange transmitters with underwater microphones beneath them that were spotted in the corners of a triangle two thousand feet on a side. It was an ambitious project—more so because the researchers were not only interested in monitoring and recording the sounds made by these whales, but they also hoped to determine which whales were vocalizing and what kind of behavior was taking place at that time. This, of course, required being able to spot the whales at the moment they were vocalizing, identifying the individuals, and somehow getting all of this information on magnetic tapes.

How complicated a feat this was can only be imagined, for the investigators soon found that the whales apparently made no special motions with their mouths or bodies while sounding off. Only by employing an airplane spotter to watch the whales' activity and radio the information to them could it be done. Meanwhile, the scientists were simultaneously trying to pinpoint a whale's sound source by measuring delays in the arrival time of the whale's voice from the three spaced underwater microphones. Payne and the others soon found that the right whales employed a complex vocabulary of strange, haunting grunts and groans. But whether or not these sounds were "meaningful" exchanges of information between the animals was unknown.

Far more obvious forms of communication were evident in the body language used by these animals. Diver-photographer Bill Curtsinger's first closeup encounter with a right whale involved a bit of body language that he had not the slightest bit of difficulty understanding. Not having dived with this species of whale before, Curtsinger was a bit apprehensive. Payne, however, assured him that it was perfectly safe to swim among the mammals. Later, he described this experience:

"Bill had no more than slipped from the boat when a whale turned to face him. At a distance of about three feet, it began slamming its head from side to side and up and down, churning and battering the shallow water into explosions of flying spray in an awesome display of raw power. Throughout his tempestuous exhibition, Bill held his ground.

"When the whale's frenzy had subsided, Bill swam slowly back to the boat, passed his camera carefully over the gunwale, pulled himself over the side, and said, 'That was fascinating.' "

During the period in which the divers were so familiar with the animals that they were often within touching distance of them, this was the only time the observers saw any kind of aggressive behavior among the whales. "I think he was just trying to tell me to scram," said Curtsinger, still unperturbed by the fact that he had stood his ground while a forty-ton animal was apparently ordering him out of the water.

For the next five years, Payne and his family researched the right whales that came to the remote Patagonia coast each winter to mate, calve, and raise their young. His interest in this particular species of whale stemmed from five years of research on humpback whales, especially their vocalizations, the song they repeated for hours on end each spring.

But what made the right whale particularly well suited to Payne's research was the individual animal's easy identification due to peculiar growths called callosities, patches of thickened white skin inches deep with a rough exterior surface; they look like large cauliflowers perched around the whale's head. The size, number, shapes, and placement of these highly visible growths made for excellent identification so that Payne could better determine which whales were vocalizing and how the various animals reacted at these times. Moreover, several of the right whales had white spots on their backs, in addition to the callosities, that helped the investigators keep track of the various animals. For example, the Paynes noted that a whale they called "Y-spot," because of the shape of its distinct marking, had a calf in 1971, then vanished from the Patagonian area for a couple years, only to return in 1974 with a new calf. From this reunion, Payne established that the pattern of markings and callosities remained constant over long periods, and it indicated that the female of the species might breed only once every three years.

After five years of observing these peaceful mammals, what amazed the Paynes was the absolute lack of aggressiveness within the herds. In fact, the only behavior that might even approach being aggressive amounted to nothing more than an occasional mild pushing or shoving between animals. And what was particularly touching was the unique relationship between the mother and her young.

"I have watched many a calf boisterously playing about its resting mother for hours at a time, sliding off her flukes, wriggling up over her back, covering her blowhole with its tail, breaching against her repeatedly, butting into her flank—all without perceptible reaction from the mother," said Payne. "When finally she does respond to the torment, it may be only to roll onto her back and embrace the infant in her armlike flippers, holding it until it calms down. It is hard to think of comparable equanimity among any other mammals, including man."

The Paynes found the desolate Patagonia coast a windy place. When the wind blew hard, the whales responded in a remarkable way. They slapped the water resoundingly with their huge flippers, turned their great tails on high, then smashed them into the sea and

hurled their forty-ton bodies completely out of the water into great
sea-smashing backward sommersaults. Compared to their normal,
quiet, slow-moving gentle manner, the great giants appeared to
have suddenly gone berserk, all, the Paynes discovered, due to an
increase in wind! Why, they wondered. What did the wind have to
do with the animals suddenly going wild, performing what the
scientists soon classified as "flipper-slapping, lobtailing, and breaching"?

The only explanation that seemed logical to the watchers was that
all three forms of violent activity might be methods of communication. Higher winds created waves and their underwater roar in the
shallow bays, in particular, increased the low-frequency noise on
the same levels used by the whales to communicate. Were the wave
sounds so disruptive to the whales' normal exchange of vocal communication that they were forced to use another means? Possibly,
too, the increased rolling seas might be all the excuse the southern
right whales needed to turn loose and, like children cavorting in the
surf, have themselves a frolicsome time playing in the water.

"When the wind blows really hard," said Payne, "there are so
many right whales splashing spray and foam that it is hard to keep
track of who is who."

Night-sailing off Bermuda, trailing a pair of hydrophones, the
Paynes recorded many hours' worth of vocalizations from the humpback whales (*Megaptera novaeanglia*), which each spring pass Bermuda on their northward migration. The whales are returning from
their annual rendezvous in the warm Caribbean waters near Puerto
Rico where, it is believed, their calves are born. From such unique
recordings as these and others made at another time of year in the
Hawaiian Islands, where the humpbacks gather again to make the
underwater world resound with their booming vocalizations, the
Paynes are largely responsible for providing us with the eerie underwater chorus whose strange reverberating acappellas have been
widely popular as the "Songs of the Humpback Whale."

After years of recording the incredibly varied and strangely plaintive vocalizations of this particular species of whale, there began the
prodigious job of analyzing these songs and comparing their tone
patterns one with another, searching for some meaning to the
unearthly submarine chorus. Hearing such sounds for the first time,
one is reminded of lowing cattle, of basso-profundo cows whose

resounding moos suggest the animals are afflicted with laryngitis, until suddenly the eerie tone slides up the scales into the higher octaves becoming reminiscent of predawn vocalizing coyotes. Other listeners have compared the sounds to a combination of oboes, bassoons, and string basses tuning up.

It would seem as if it might take several lifetimes to sort out and make sense of the sounds these alien species broadcast in their watery world. But after listening to many recordings and analyzing them with the help of Scott McVay at Princeton University, the invaluable records that created a "voiceprint" of the humpback's song revealed that the giant mammals were making a regular sequence of repeated sounds that might vary considerably in pitch, but maintained the same theme. Only the variations changed. Surprisingly, the song of the humpback whales was so similar to a song a bird might sing that, when the scientists speeded up the sound tapes about fourteen times faster than normal, the whales did indeed sound like birds!

Once the investigators could visually chart the whales' song, the Paynes saw that their hydrophones had picked up solos, duets, trios, and full choruses, with the whales intermingling their parts as if a dozen members of a choir were singing the same song, but not one was in unison with another. Moreover, some of these seagoing choir members were singing variations on the theme, sticking largely to the main melody, but adding new elements as they went along. This is an astonishing revelation, for no animal other than man has ever been known to be capable of this kind of complex behavior. Experts are at a loss to find an explanation for it.

Over the years, the Paynes have been amazed at how different the songs are. Songs taped from two different years—1964 and 1969—for example, sounded as different to the researchers as do Beethoven and the Beatles. When tapes made by other investigators were combined with theirs, the Paynes had a representational record of humpback vocalizations that spanned twenty years. Now, perhaps something significant could be learned from the analysis of this vast record. And what they learned was as astonishing as the variations of the songs themselves. It seemed the whales all sang the same song, which changed from year to year, evolving from common themes that existed the year before.

Humpbacks winter in the Hawaiian Islands, where they vocalize the current year's song. It is doubtful that the whale populations of Hawaii and Bermuda have any contact with each other whatsoever, yet the Paynes found striking similarities in the manner in which the different songs are composed by these two groups.

"Each song, for example, is composed of about six themes—passages with several identical or slowly changing phrases in them," said Payne. "Each phrase contains from two to five sounds. In any one song the themes always follow the same order, though one or more themes may be absent. The remaining ones are always given in predictable sequence."

Based on this evidence, it would appear that the whales in both parts of the world inherited the same rules of composition within which they improvise variations. Has this unique ability been passed genetically from generation to generation of humpback whales, or is it a learned ability?

How good is a whale's memory, wondered the scientists. Since each year the whales sing a different song, one amounting to a variation on an old song, could it be that they remembered only fragments of the old song from one year to the next and were simply embellishing what little they recalled? After all, it was known that the whales did not sing all the time, but rather for only a few months out of the year, the time spent in their wintering area in the Hawaiian Islands. They never sang at their summer feeding grounds. So it was perfectly logical that, unless they possessed the elephant's mythical memory, they might have forgotten much of their old song. Hence the new variation.

To test this theory, the Paynes initiated a prolonged observation and recording session during one season with the humpbacks off Maui in the Hawaiian Islands. The six-month effort brought new revelations. Analysis of the humpback's vocalizations in their winter quarters revealed that the whales arriving in the islands had not forgotten their previous season's song, because the Paynes recorded them singing it when they returned to Maui. The changes and variations on that theme occurred there as the season progressed. This was the breeding season for the whales, and it appeared that they were capable of remembering the main theme from one season to the next without change until that particular moment.

Another intriguing fact revealed through analysis of the recordings was that whales apparently were singing the old fragments of the song, those they knew well from the previous year, much more rapidly than the variations they had added. As these facts became evident to the scientists, they gained greater respect for the whales' intelligence and their mysterious songs, which appeared much more complicated than anyone had originally imagined.

The Paynes' observations revealed an even more intriguing revelation about the humpback's ingenuity—the uncanny and seemingly cleverly reasoned "spinning of bubble nets." This unusual behavior was noticed by Charles and Virginia Jurasz and their family, who have been observing humpback behavior in the icy waters of Juneau and Glacier Bay, Alaska, for over twelve years. Not only are the members of the Jurasz family whale watchers, but they are whale watcher watchers, noting the reaction of the wintering humpbacks to the presence of the numerous private vessels that annually bring hoards of people to these waters to observe the antics of the feeding whales. For years, the Juraszes have kept records of the visiting whales; they have put together an impressive "Fluke File," an identification chart for each whale they sight, naming it after some characteristic marking. For example, one is called "White-Eyes" because of a set of white eyelike patches on its tail. Another is named "Spot" for a prominent fist-sized white mark on the underside of its tail fluke. Then there is "Dot-Dash," an animal with a mark that looks like an inverted exclamation point.

As far as the Juraszes have observed, the Glacier Bay humpbacks are not the same whales that summer in the Hawaiian Islands, at least no reliable verification of their being so has yet been obtained. Nor do these humpbacks sing songs. They spend most of their time eating and sleeping, feeding on the wealth of food concentrated in these cold waters, especially the krill, a kind of small shrimplike crustacean.

The normal method by which these whales feed is to open their toothless maws, then plow forward through the seas, engulfing in a single gulp small fish, crabs, krill, and huge quantities of almost microscopic-sized aquatic animal life. These are strained from the seawater through the sievelike baleen plates of the whales' upper jaws. But since the waters of the bay cannot concentrate this food in the kind of numbers whales prefer, some of the larger, choicer morsels such as schools of krill require more than just a casual sieving of the seas to satisfy their large appetites. So the whales have devised a unique way of concentrating their catches. They dive deep beneath schools of their prey, and as they swim toward the surface in a spiral, they emit clouds of bubbles that rise rapidly in a kind of silvery curtain surrounding the schools of krill and momentarily at least driving them into the middle of this "bubble net." Before the krill realize they are being fooled, the humpback rises swiftly through the center of the column with his maw agape and consumes the entire school. This behavior has been observed so often that whale watchers feel sure it is an intentional, intelligent hunting tactic.

When Roger Payne heard about the humpbacks' bubble nets, he was anxious to record the sounds that occur when the whales blow these bubbles. Payne already knew that humpbacks off Maui had been seen vocalizing. They were, however, not only usually motionless, but they also emitted sounds without ejecting any bubbles. Scientists believe they do this by squeezing masses of air through different parts of their anatomy. When Payne learned that the Glacier Bay humpbacks were actually making bubbles, he hurried to join the Juraszes and record any sounds associated with the phenomenon.

Later, analyzing the noise patterns of the bubble-blowing whales, Payne found evidence that they could control the size of the bubbles and thus of the "mesh" of their "nets." Though it appeared that this behavior was in no way a social or communicative activity, it once again suggested that the whales had a higher intelligence than originally suspected, one that at least correlated with that of an animal intelligent enough to set a trap deliberately.

Accompanying the Paynes part of the time during their study of the Hawaiian Island humpbacks were renowned underwater diver-photographers Al Giddings and Chuck Nicklin, and marine biologist Sylvia Earle. While the Paynes recorded the humpback whale songs, the others went underwater to record the animals' activities on film. They were probably the first to photograph a singing whale. Said biologist Earle, "Underwater, the song was so intense that we could feel the sound as the air spaces in our heads and bodies resonated." When he heard the song at close range, photographer Giddings said, "The sound was incredible, like drums on my chest."

Humans can detect the presence of sound underwater, but not its direction. In water, sounds are interpreted by man as being omnidirectional, seeming to come from all around at the same time. This is a definite disadvantage to divers, who become separated under conditions of low visibility. If one diver tries to home in on the other by rapping on his metal air tank with his knife, the latter has no idea from which direction the sound comes. The reason for this, say the experts, is that underwater sound vibrations are received by the entire skull and not only by our two ears as they would be in the air.

Evolutionists tell us that, as land animals evolved into water animals, their land ears also evolved into organs of little use. Therefore, the whales gradually lost their land ears until today their ear canals are no larger than pinholes and in some species they are plugged with hard wax.*

For underwater creatures such as the cetaceans, other methods of hearing evolved. Sound is transmitted to these aquatic species through material less dense than water—certain blubber fats, fatty deposits inside the whale's jawbone and especially in an area of the domed forehead that contains a fatty deposit called the "melon." It not only can receive sound and transmit it to the cetacean's inner ear, but we now know that this melon is also used by the mammal to broadcast sound. Though we do not yet understand all of the mechanics involved in this unique feat, we know that whales can both broadcast and receive sound waves simultaneously.

Investigators studying underwater sound as it is emitted by whales have classified these sounds into two categories: pure tones such as those made during the songs of the humpback whales, and pulsed tones at one frequency that whales use for echolocation and at another frequency for communication. Of course, cetaceans can create other sounds, explosive sounds, merely by hurling their bodies out of the water to breach, by slapping the water with their flukes, or by snapping their jaws shut. Each disturbance creates a

---

*In baleen whales each canal contains wax ear plugs layered light and dark by annual deposits of wax. Since the rings resemble growth rings in a tree, some scientists have found that counting the layers of wax is one way of determining the life span of these whales. The waxy ear plug of a big fin whale contained eighty layers, suggesting that these large whales probably live as long as man.

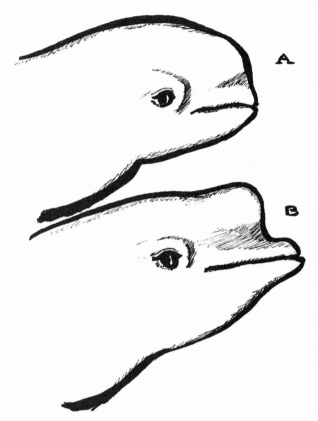

When not sounding off underwater, the beluga or white whale's forehead appears as in (A), but when communicating with each other, the whale contracts muscles above its jaw and bulges out the upper part of its head as a sound chamber that beams out the signals. So loquacious is the species that British sailors call them the "sea canaries."

different kind of sound that is undoubtedly significant to other members of the whale family.

Since sound travels through water at about one mile per second, communication in this medium is comparatively rapid and capable of reaching long distances. The underwater medium is so ideal for this kind of transmission that sound literally bounces around underwater like a rubber ball, caroming off the underside of the sea's surface, deflecting off the bottom, bouncing back from a wall of suspended plankton or sediment, or simply ricocheting off into watery space. How far it may go depends upon a number of variables, such as water temperature, depth, pressure, kinds of suspensions in the medium, or salinity, not to mention the initial intensity and frequency of the primary transmission.

Whale sounds have been recorded fifteen miles away from the only animal in sight; but it is believed that the sounds from some of the larger whales may travel hundreds of miles. If this is true, cetaceans may be able to keep in contact with one another at these incredible distances.

Two major classes of whales exist: the larger, slower-moving baleen whales that lack teeth, the filter feeders such as the humpback and the blue whales, and the smaller-toothed whales such as orcas and dolphins. Both of these classes of whales produce significantly different kinds of sounds—the baleen whales emitting the moaning, low-pitched groans and pulses; the toothed whales a complex repertoire of high-frequency sounds varying in pitch and repetition rate. Some of these latter sounds are known to be used as the animal's sonar, to help it echolocate. Within these two major classifications of whales, the baleen and the toothed, the various kinds of animals make largely different kinds of sounds. While the humpback whales may seem to be singing their hearts out, gray whales, for example, appear to have little to vocalize about. And then, the white whale or beluga *(Delphinapterus leucas)* is so vocal it has been nicknamed the "sea canary."

Man has been listening to leviathans for centuries. Even the ancients knew that these sea creatures were not mute. But modern man has had difficulty defining what kind of sounds these species make. For example, as early as 1840 the sperm whale was reported to make creaking sounds. In 1957, two scientists increased that sound category to "a muffled smashing noise, a low-pitched groan, a rusty hinge, crackling sound, and clicks." Four years later, another scientific observer heard a harpooned sperm whale clicking beside the catch boat. In 1962, a phonograph record of sperm-whale sounds included the statement that clicking was the only sound ever heard from this species. Four years later, however, we are told that some of the sounds, including "muffled smashing noises and a grating sort of groan," were merely click sounds being distorted by the mechanical recorder. Today, scientists have "successfully recorded commu-

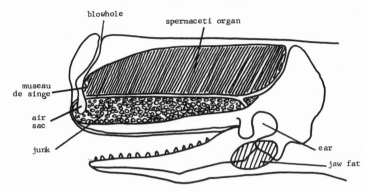

The sperm whale's sound-making organ produces rapid clicking ranging from one to 55 clicks *a second,* believed to travel hundreds of miles underwater. The sound originates by clapping together the lips of the museau de singe. The sound travels through the oil-filled spermaceti organ, ricocheting off skull bones and air sacs until it escapes through a "window" indentation in the upper jaw. Whalers called the cheaper fatty portions of the sperm whale's head "junk." The more valuable spermaceti organ compares to a dolphin's melon, used to focus the sound, produced largely, probably, for echolocation. After the *Porpoise Watcher,* New York: Norton, 1974.

nication 'signing' from the sperm whale." In other words, it appears that these huge leviathans can broadcast an identification signal that tells the other herd members who they are.

Here is an example of how we interpret some of the other sounds that the different whale species make. We find that the echolocation and communication signals of the gray whale *(E. gibbosus)* consist of clicks, grunts, mumbles, groans, rasps, and a loud, resonant bong. The Byrde's whale *(B. edeni)* communicates in moans, trills, and sighs; while the fin or finback whale *(B. physalus)* moans, screams, emits high-pitched whistles, cries, chirps, and squeaks. The blue whale *(B. musculus),* the largest animal in the world, attaining lengths up to one hundred feet and weighing in at around two hundred tons, primarily clicks or broadcasts groups of clicks that merge together to form buzzes and rasps, with some low-frequency tones included among them.

According to biologist and research oceanographer Theodore J. Walker, the sperm whale *(Physeter catodon),* the species that Moby Dick belonged to, sends out a loud clacking "signal roughly equal in decibels to the noise heard by a man twenty feet directly behind a jet engine at wide open throttle." According to Victor B. Scheffer, chairman of the U.S. Marine Mammal Commission from 1973 to 1976, when Navy zoologist William C. Cummings and Paul O. Thompson hooked up electronic instruments capable of measuring the intensity of a blue whale's vocalizations off the coast of South America, they were astonished to find that it was "the most powerful, sustained sound from any living source and, because of its pitch, could surely be detected for hundreds of miles. They rated it at 188 decibels, comparable to the same overall noise level of a U.S. Navy cruiser traveling at normal speed."

So it seems that, among the cetaceans, as among humans, some are soft-spoken while others are veritable loud mouths. This is as true among the baleen whales as it is among the dolphins, members of the toothed-whales class. *Tursiops truncatus*, the bottlenose dolphin of "Flipper" fame, is extraordinarily loquacious, while his relative, the long-beaked blind river dolphin, Ganges susu *(Platanista gangetica)* of India, seems to have little to say. It clicks a lot, however, supposedly to find its way around the always-muddy Ganges River. It uses its clicks to judge its distance from unseen obstacles the same way a blind man listens to the echoing taps of his cane to "see" with sound.

Up until now, scientists eavesdropping on whales with their sophisticated listening and recording devices have catalogued a galaxy of varied sounds that have originated from different species. Though we may be able to listen to these oceangoing aliens, record their sounds, and chart their vocalizations in the minutest detail through the use of computer enhancement, analyzing each and every squeak or groan they produce, we are still almost as ignorant as ever of what these sounds mean, if indeed they mean anything. Studying spectrogram charts detailing every nuance of their emissions, investigators have found these sound pictures as complex and difficult to comprehend as is the picture language of the Mayas. But this does not deter the ongoing investigation. Modern-day Champollions are trying every way possible to "break the code," to learn what meaning exists in these vocalizations. When the breakthrough comes, as come it must, one hopes that with this understanding man has the intelligence to avoid abusing his new knowledge as he has earlier abused the entire race of animals. Though they have always shared our world with us, whales are an alien life, one we have until now hunted and practically decimated to the point of extinction, one we know about only from their dead carcasses, and one we have never considered intelligent enough to be worthy of any further attention. Even now, as so-called modern man enters the space age, one wonders if he is mature enough to be able to cope with even the present knowledge we have acquired about these unique animals. As marine biologist Sylvia Earle so aptly put it, "We dream of communicating with intelligent life in space. But I wonder how this goal can be attained when we have not yet achieved peaceful rapport with whales, gentle earth creatures that even share our mammalian heritage."

8

┣┉┿┉┿┉┿┉┿┉╋┉╋┉┿┉┿┉┿┉┿┉┿┉┿┉┿┉┿┉┿┉┿┉┿┉┿┉┿┉┿┉┿┉┿┉

Mammals That See Sound

"To enter into the perceptual world of whales and dolphins, you would have to change your primary sense from sight to sound," wrote naturalist Peter Warshall. "Your brain would process, synthesize, and store sound pictures rather than visual images. Individuals and other creatures would be recognized either by the sounds they made or by the echoes they returned from the sounds *you* made. Your sense of neighborhood, of where you are and who you are with, would be a sound sense."

And this is the primary sense employed by whales and dolphins. These remarkable marine mammals have evolved sound senses capable of performing feats that we human animals of "higher" intelligence can only marvel at. Here is a sample: In a swimming pool–sized body of water, a *blindfolded* dolphin is asked to find a target—a half-inch-long vitamin capsule. Without hesitation and without visual aids, the dolphin swims directly to the invisible target.

In another instance, a dolphin is asked to differentiate between two pieces of metal painted the same color, one a square of copper and the other a square of brass. Unerringly, the dolphin easily distinguishes the copper square from the brass square even though both targets look exactly alike.

Even more uncanny, a blindfolded dolphin can instantly "see" you in the water with him; he can also tell things about your internal makeup that only a doctor might otherwise determine with an X-ray.

How is it possible for these marine animals to perform such astounding feats? This is one of the questions cetacean researcher

Kenneth Norris asked himself upon observing how well dolphins in captivity were able to solve such seemingly unsolvable problems.

It was as if the animals were capable of reading the experimenter's mind, as if they possessed some kind of super sense that replaced their sight so that even though blindfolded they could "see" the tiny targets researchers were asking them to identify.

Thanks to the continued efforts of such early researchers as Norris, we now know that dolphins do indeed possess a super sense—a sound sense—that among other things allows them to pinpoint targets through echolocation with their cetacean sonar (SOund Navigation and Ranging) system, even when blindfolded.

One of the first, if not the first, to realize that dolphins use some kind of extraordinary sensing system was Arthur McBride, the first curator of Marine Studios in Florida. McBride captured wild dolphins and displayed them to the public there. Normally his catch boats cornered the dolphins in murky canals and bays and netted them. But what surprised McBride was how easily the dolphins were able to avoid his fine-meshed nets. In muddy water with mere inches of visibility, the dolphins often turned away from his nets long before they reached them. Yet they could neither see nor hear the nets. How did they do it, McBride wondered.

Was it possible that they were using a system similar to that used by bats, and were able to send out sound signals that bounced back to them from off the nets? McBride believed that this was the case. He theorized that bubbles forming on the fine mesh of his nets acted like a solid barrier, bouncing the sound signals back in warning to the dolphins.

If this was true, he could surely check his theory. McBride went after the dolphins with a much larger mesh net, and this time he was successful in capturing them.

McBride was not about to reveal his new knowledge to any potential competitors, and his early awareness of dolphin sonar and his method for foiling it were not revealed until many years later, after his death. When researchers William Schevill and Barbara Lawrence were trying to learn more about how dolphins navigated, they found and published McBride's early field notes. Later, working with independent investigator Winthrop Kellogg at Marineland in Florida, the researchers performed test after test with the captive

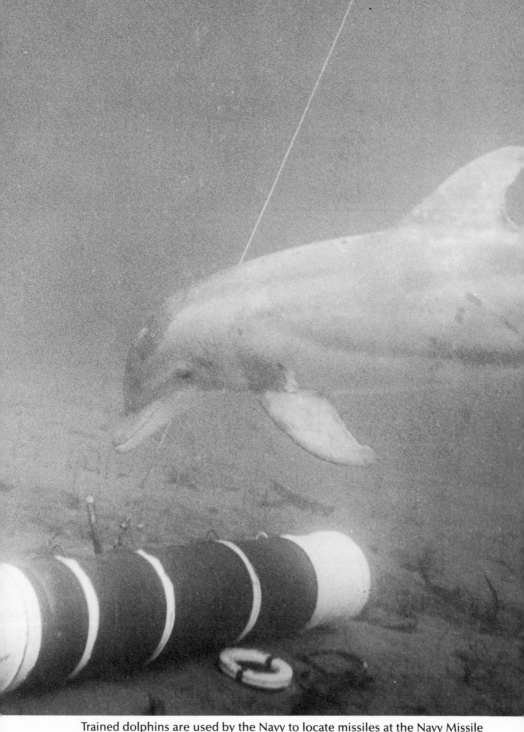

Trained dolphins are used by the Navy to locate missiles at the Navy Missile Center, and plant a marker buoy at the site. Official photograph U.S. Navy.

dolphins at that aquarium and subsequently provided the basis for additional understanding about this unique sense.

It took long enough to learn exactly what the mammals were doing that enabled them to perform such marvelous feats of "seeing with sonar." But it was far more difficult for them to determine exactly *how* the animals did it. What the researchers found was that all the toothed whales and some of the baleen whales use this sound-navigation system. Somehow—and it was not completely understood how—these animals emitted a powerful *click* that swept through the water until it encountered an object denser than the water. Then, an echo of that encounter bounced back to the dolphin and was simultaneously processed by the animal. It told him how far away the object was by the interval between the click and the returning echo. In its simplest terms, this was the basic process. But only high-speed repetition of this clicking and interpretation of returning echoes provided the dolphin with the information he sought.

In their experiments, Schevill and Lawrence found that their test animal could repeatedly echolocate a target fish dropped anywhere in its tank in this manner: the dolphin emitted a series of intense sonar clicks of about twenty to thirty per second, almost simultaneously receiving the echo—the reflected wave altered now in character. As the dolphin turned toward the target, emitting more rapid clicks and noting the changes in the returning echos, these sound-wave modulations told it things about the target's size, speed, location, and makeup of the target fish itself. As its clicking continued, and it closed in on the target, the dolphin had a complete readout of the situation, right up to the moment it devoured the prey.

As we have discussed earlier with other whales, these animals are able to broadcast a beam of sonic signals with their melon, the oil-filled dome on their forehead, then pick up the returning signal not only with the same organ, but with a thin oil located in the dolphin's normally out-thrust jaw. Both areas then direct the sound inward through a so-called "acoustic window," a thin, hollow shell of bone covering an oval area in the rear part of the dolphin's jaw near the skull. And from there it is transmitted to the mammal's inner ear.

Tuffy, one of the Navy's first dolphins, carries a life-line to a diver in a simulated rescue mission during Sealab. Official photograph U.S. Navy.

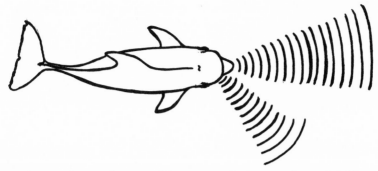

Using their melon to shape and focus sounds they produce, dolphins can direct one series of echolocating signals forward in an arc normally covering an area measuring ten degrees on each side of a midline down the dolphin's body; it can simultaneously shoot out a probing side beam. To cover a wider range, the dolphin will swing its head, sweeping the projected beam back and forth.

Researchers learned that the fatty tissue in the dolphin's forehead, the melon, was also capable of focusing this sound—literally beaming it out in different directions, the same way you might beam a spotlight into the darkness to illuminate some object. But astonishingly enough, with the dolphin these echolocation patterns were not only broadcast out at a normal ten degree arc from either side of a midline down the animal's head, but the animal also could shape the beam, make it broader or narrower, or even project a second beam at an angle to the first.

For wide coverage the mammal had only to swing its head from side to side. Moreover, the beam could be fine-tuned. A high-pitched note made up of shortly spaced high-frequency sound waves traveled a shorter distance, but it created a far more detailed sound picture for the dolphin. Or, the dolphin could use a low-frequency click resulting in a lower note that traveled a far greater distance but lacked the sharply defined return sound image.

How do we know all these things? How do we know for example, that when a dolphin echolocates he broadcasts the sound through his melon, rather than say, his larynx? The answers to these and to all such questions were learned only through the work of such people as Kenneth Norris, who ran many painstaking tests on dolphins in captivity. Norris, who pioneered work in dolphin sonar, performed much of his research on a dolphin named Zippy. To demonstrate that this dolphin's sonar definitely did not originate in the larynx as was earlier believed, Norris blindfolded Zippy using soft rubber suction cups over the dolphin's eyes. Then he studied how the animal could locate food without being able to see it visually.

In his experiments, Norris learned that, if food was anywhere above Zippy's beak, she always found it. This indicated that the sounds were emanating from somewhere above, in the upper head. To localize their source even further, researchers found, with the aid of hydrophones, that these high-frequency click "trains" (or high-speed series of clicks) reached their greatest intensity just ahead of and over her snout.

This indicated that the melon was most probably the source of the emanations. Norris and his assistants fashioned a kind of helmet that would mask that part of the dolphin's head and therefore cancel out the transmissions. But at this point Zippy stopped cooperating. Each time the scientists tried to put on the mask, she shook her head violently and flipped it off. But to show them that there were no hard feelings, she would pick it up in her teeth and obligingly hand it back to the experimenters.

Such were the kinds of setbacks Norris and all such investigators were constantly confronted with in dealing with often temperamental test animals. But somehow, with much patience, the experiments continued, and gradually the bits of evidence mounted until the scientists were able to come to more positive conclusions about how these animals achieved their marvelous feats.

Describing the dolphin's capability to manipulate a second sound beam, Warshall learned that wild dolphins in the open ocean use multiple sound frequencies to listen acoustically at both close and distant ranges. What was so remarkable was how quickly the mammals could do it.

"The dolphins can switch frequencies in less than one-thousandth of a second," said Warshall. "But whether they use high or low frequencies, they can sputter out clicks with incredible rapidity— up to three hundred per second—and still interpret this fast-fire sound echo."

Soft rubber eye cups are used to blindfold a dolphin so scientists can test her ability to negotiate an underwater obstacle course by using only her sonar. Courtesy of Marineland of Florida.

Water conditions often determine the different kinds of sonar employed by the cetaceans, believes Warshall. For example, the sperm whale, living in clear deep seas, probably resorts to the low-frequency signal necessary for that animal's far-reaching deep-water requirements. In comparison, certain species of muddy-river dolphins, which live in murky water most of their lives, sound off with faint but fast-clicking series apparently more useful to them in their shallower, muddier environment.

Analyzing the characteristics of this underwater sound broadcast by cetaceans, investigator Robert McNally observed that echoloca-

Testing how deep a dolphin can dive, this wheel device suspended from a cable contains a mercury switch and buzzer. The buzzer signaled the dolphin, Tuffy, to dive down and tilt the device, turning off the buzzer and telling topside observers that Tuffy had gone to the desired test depth. Official U.S. Navy photograph.

tion sound waves operate much in the same way as a beam of light. When the light changes by passing through or reflecting off some object, our eye detects this different characteristic and from it our brain interprets the image. However, there are differences, said McNally. "For one thing, [sound] penetrates much better, bending around corners and passing through things."

Mediums of different densities reflect different sound echoes; therefore, said Warshall, "If a human diver jumps into the water with a dolphin, the dolphin can 'see' inside the diver into the air passages of his lungs and respiratory system. This is because sonar sight penetrates materials that are apparently the same density as the water—like human flesh—and returns different echoes from objects with different densities. The greater the difference in density, the more easily sonar can discriminate. In the case of the diver,

During deep-diving tests with Tuffy, this device was used to obtain a sample of the dolphin's exhaled air after the dive. Here, Tuffy presses a buzzer to tell trainers he had performed as requested; the air sample is being deposited in the funnel-shaped cone over his blowhole. Official photograph U.S. Navy.

his lungs show a greater contrast to the water than his wetsuit. To the dolphin, the diver might look like an X-ray photograph of the human body."

Once the researchers saw and marveled over the acoustical feats of captive cetaceans, they began probing at the mechanics of this mysterious system. These experiments were funded—and still are being funded—by the U.S. Navy, which understandably would like to know precisely how these animals produce and use this phenomenal ranging system. Indeed, the navies of the world would like to know all the secrets of cetacean sonar. One can well imagine what an incredibly sophisticated sensing system could thus be acquired if it were possible. So far, however, man has learned enough to be able to duplicate only a few of the minor miracles nature has bequeathed to these aquatic mammals.

One of the biggest mysteries that confronted researchers was how an animal that lacked vocal cords, and therefore had no voice, could still manage to send out high- or low-frequency rapid-fire sound signals. The secret seemed to lie in an area at the top of the dolphin's head; it appeared to have something to do with air from the animal's lungs being forced into a lipped air sac located between the dolphin's melon and his blowhole, where he breathed.

In an effort to unravel the mystery at the University of California at Santa Cruz, Kenneth Norris X-rayed dolphins that were emitting click signals. In the resulting films, he noticed that a nasal plug, a lip

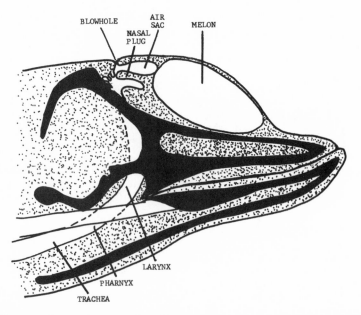

This diagram shows the sound-making organs of the bottlenose dolphin. When vocalizing underwater, blowhole is kept closed. Air from the lungs is forced into the air sac, distending it. This air is then released back into the lungs causing the nasal plug to vibrate against its corresponding lip. As it is reflected by bones of the jaw and brain case, this sound is shaped and focused by the dolphin's oil-filled melon, then beamed out in a ten-degree arc on each side of a midline through the dolphin. Echoes of those signals are picked up again through the mammal's melon and through thin oil in the jawbone where they are transmitted to the cetacean's inner ear. Dolphins can broadcast and receive signals simultaneously.

of flesh just inside the blowhole, vibrated during the process. Norris figured that the click sounds were produced by shifting air back and forth between it and the pharynx through the nasal passages, causing the nasal plug to vibrate against nasal bones. A similar series of vibrating sounds, using the same mechanical principal of one object rubbing against another, can be produced as follows: at some restaurants and drugstores, coffee is served in a cone-shaped paper cup supported by an outside plastic frame containing the handle. With a full cup of coffee and your fingers curled through this handle, place your thumb on the edge of the cardboard cup rim and move your thumb back and forth, rubbing the rim just hard enough to make it vibrate. Notice how easy it is to create a whole series of "click trains" by hardly any movement of your thumb against this surface. Now, think of the dolphin "thumbing" the firm lip of flesh just inside its blowhole with its nasal plug and you can understand how its high-speed click trains are most likely produced, if indeed it is done this way.

French bioacoustician Guy Busnel has a slightly different theory of how these sounds are produced. Busnel believes that cetaceans bring them about by pinching air as it escapes through their nasal passages with two or three organs including the nasal plugs. In effect, he believes they are making the sound frequencies in the same way a child might make air escaping from a balloon literally "sing" by pinching it as it escapes through the neck.

Whichever theory is correct, we will probably never entirely explain how these animals have such incredibly precise control over so wide a range of different clicking characteristics. For example, it may take more than human intelligence to comprehend how dolphins can perform such feats as varying the click rates at will, broadcasting up to one thousand or more sounds *per second,* changing the frequency of the clicks upward or downward while maintaining what sound experts call "the energy peak." Moreover, while broadcasting broad band clicks, dolphins can simultaneously fire off narrow band signals called "whistles," possibly by producing different sounds at the same time in each nasal passage.

Complex and wonderful as the dolphin's sonar system is, man is slowly but surely closing the gap in understanding how it all works and how we may duplicate some of their feats on a smaller scale. As a result of extensive inquiries into the mechanics of cetacean sonar, scientists today are using side-scan sonar systems actually to draw three-dimensional pictures of such things as underwater shipwrecks in the deepest part of the ocean. Or, they can actually look through the bottom of the ocean, penetrating the less dense material to bounce signals back from denser objects that may lie hidden beneath the substrate.

"Our present electronic equipment can't match a dolphin's abilities—not by a factor of ten," said Richard Soule, director of the Biosystems Division of the U.S. Naval Oceans Systems Center Laboratory in Hawaii. The dolphins are still ahead of us, but in this computer age, we are fast catching up. Many of the dolphin's high-frequency echolocation signals are too high for us to hear. But, Navy scientists can now electronically reproduce those sounds, broadcast them underwater, and with the aid of a computer, lower the sounds into a frequency range that humans can hear. This has enabled blindfolded divers to tell the difference between two targets with the same kind of dolphinlike ease that once baffled researchers of but a few decades ago.

Naturally, the Navy hopes eventually to perfect this experiment into a kind of miniaturized biosonic ranging system enabling divers to see where normal sight is impossible, as a kind of guidance system not only for humans, but for such underwater ordinance as sonar-equipped target-seeking torpedoes. We are not alone in this area of research. The Soviet Union has been studying dolphins since 1967 and their findings are most likely being directed toward the same avenues of use as are ours. Not all of it, however, need be solely for military application. By the early 1970s, an echolocating device was being researched by the Russians for use by blind people. One example was the ultrasonic "Orientir" device developed by Soviet engineers. This instrument had an operational range of about thirty feet. Completely transistorized, it weighed 230 grams (8.05 ounces). A blind person carried it like a flashlight, the small transmitter beaming out directional supersonic signals. Amplified, the echoes of these signals were converted into sound waves that changed in tone in the user's earphones. From the pitch of a sound a blind person could judge the distance to an object, and from the timbre of that signal, could determine the character of the object.

A staff member of the Naval Missile Marine Biology Facility at Point Mugu, California, conducts a test with a dolphin. Navy scientists must understand dolphin physiology to use most effectively the mammals in their "Man Under the Sea Programs." Official photograph U.S. Navy.

Change in tone indicated the object was moving, approaching, or retreating. This unique instrument is based on the same kind of echolocation procedure practiced by animals with this capability. Miniature electronic homing devices are already on the market for divers and it is just a matter of time before echolocation instruments may be just as readily available.

While naval research scientists try to imitate the dolphin's super-sonar system, other investigators seek to comprehend another puzzling sound made by the mammals—a pure tone emission that researchers call "whistles," which scribe sonagrams showing trills, vibratos, and glissandos, sound pictures that look astonishingly like sonagrams made of bird calls. Researchers believe these are sound signals employed by the dolphins to communicate certain kinds of information to each other. Scientific eavesdroppers counted some two thousand different whistles emanating from dolphins in captivity.

"On that basis," said Jacques Cousteau, "one might conclude that the language of the dolphins is composed of two thousand sounds—or we might say, two thousand 'words.' "

But those who listened and recorded soon learned that these sounds were not always the same, nor to the observers did they seem to have any meaning to the mammals. Only a few seemed meaningful and as yet no one has been able to correlate certain identical whistle sounds with corresponding identical actions.

Dolphins are also capable of emitting a squawk of alarm, a well-delivered Bronx cheer, which leaves no doubt whatsoever in a listener's mind about the feeling the animals are expressing, because they are so similar to familiar vocalizations made by humans. Indeed, those who work with dolphins in oceanariums know only too well the mammals' great capacity for mimicking sounds, both human and nonhuman, that may interest them. This, however, is a completely different kind of response from the whistle signals. Some nonscientific speculations have suggested that quite possibly the whistle signals used by these mammals were in some way similar to the whistle language used in some parts of the world by people living in mountainous areas where distance and the sound of the wind normally drown out a verbal exchange.

The idea that dolphins may be exchanging highly codified information through their whistles strongly appeals to those who believe

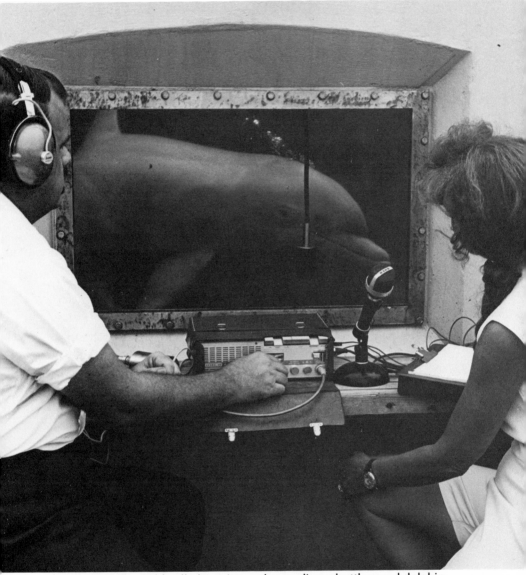

David and Melba Caldwell observing and recording a bottlenosed dolphin. Note air bubbles escaping from the dolphin's blowhole indicating that he is sounding off. The Caldwells established that the mammals emitted "signature whistles," individual clicking signals that the animals could easily recognize from each other. Courtesy Marineland of Florida.

dolphins capable of exchanging complex and meaningful information. As yet, however, no one has discovered the key to this language, and there are those who discount the extent of the use of these whistles or conversations.

Two such scientists are David and Melba Caldwell, highly respected authorities in the field of cetacean research, much of it in the area of interspecies communication of dolphins. Although their work has taken them to all the major oceanariums in the country, most of it has been done at their Biocommunication and Marine Mammal Research Facility at the Cornelius Vanderbilt Whitney Marine Research Laboratory of the University of Florida. This facility, located at Marineland of Florida near St. Augustine, has enabled the Caldwells to carry on critical studies of the dolphins' communication capabilities. Based on these experiments, the Caldwells believe that the dolphins' "whistles" really lack much variety when coming from a single animal, no matter what stimulated it to sound off in the first place. Thus, if the mammals are indeed "talking," say the Caldwells, their conversations are severely limited to an extremely concise vocabulary.

Following long and intense sound-recording sessions with captive dolphins, after which signal comparisons were made through the use of sonagrams, the Caldwells concluded that each dolphin produces a characteristic pattern of whistle all its own. He is, in effect, whistling his own tune, one the Caldwells believe is a "whistle signature." It identifies the whistler to other dolphins in the group. No matter what happens to that dolphin, whether he is in or out of the water, swimming with other members of his group or not, whether he is emotionally upset, or actually experiencing pain, the individual dolphin's signature whistle varies hardly at all. And what variations might occur are simply those having to do with intensity, slight frequency changes, how fast the signal is emitted, and how loud and how often it may be given. But it is almost always the exact same signature whistle.

The Caldwells certainly believe that these whistles impart information to other dolphins within hearing, but it is in the caliber of that information that they differ most from those individuals supporting the larger-vocabulary theory. The Caldwells believe that one dolphin can identify itself to the others by its signature whistle as well as communicate its level of excitement and probably other basic bits of information such as its need for assistance, the fact that it does not feel well, or that it is emotionally upset.

When the bottlenose dolphin emits two kinds of sound signals—the pulsed sounds comprised of what we might describe as barks, squawks, yelps, and squeaks; and the pure tone whistle signals—this combination probably signs to other members of the herd, "This is me, old Scartail, and I'm annoyed about something."

The Caldwells believe that, upon receiving such a signal, other members of the dolphin group will react according to their past experiences with this particular member of their clan. If old Scartail happens to be a dominant male and he announces that he is annoyed or angry, chances are the herd's reaction will be to stay out of his way.

Conversely, say the Caldwells, a whistling infant dolphin trying out his signature whistle for the first time may attract considerable interest and sometimes adult assistance if necessary. From the juveniles sounding off, the researchers found that the young possess equally individualistic whistle signals, but they are much simpler in form than are those of the adults. Apparently, these somewhat ragged phonations are practiced repeatedly until they are given with more assured polish, much the way a young musician might practice a certain musical passage until he becomes proficient at it.

Essentially, all dolphins whistle a characteristic song, much the way certain bird species produce certain characteristic calls. The Caldwells found, however, that as the dolphins mature, they begin to add slight variations to their signature whistles. These variations may contain additional elements of information for other dolphins, but as yet investigators have been unable to correlate these sound signals with either emotional or physiological dolphin behavior. Said the Caldwells, "We have found absolutely no evidence that the whistle has a true language content."

Still, there was enough evidence to indicate that certain information was being communicated in this manner. To test their theory that each dolphin possesses an individual whistle signature immediately recognizable by all the others, the Caldwells set up a series of revealing experiments. First, they tape-recorded the signature whistles of individual dolphins as they vocalized under a variety of conditions. Then they selected a test animal they knew was as familiar with these signals as the researchers were.

This dolphin was trained to push a target paddle each time it heard a whistle signal it recognized. Then, the researchers mixed these well-known signature whistles with those of a wide variety of other dolphins so that the signals were sprinkled indiscriminately among others in the recording.

When this tape was played back to the test dolphin, not only could it unerringly pick out each of the whistle signatures that it knew (remaining silent on those it did not know) but in subsequent experiments it demonstrated for the researchers that it could duplicate the feat eight months later with equal success, indicating that it had remembered the signatures perfectly.

Any question in the Caldwells' minds that by using their sound-analyzing equipment they could tell the difference between the dolphins' individual whistle signatures better than a dolphin could was quickly dispelled when they found that their test animal needed no longer than half a second to identify individual whistle signatures.

Observers have noticed that, when dolphins whistle underwater, the sound is accompanied by a stream of air bubbles escaping from the blowhole. In one of the early experiments to understand the mystery of dolphin vocalization, a mother and her female offspring were temporarily separated into two tanks with a shallow water channel connecting them. During this separation, the young dolphin whistled persistently, while the mother answered frequently and remained close to the gate of the connecting channel. When the channel was drained, breaking the continuity of the sound-conducting water, the mother seemed to lose interest in communicating with her offspring, and left the gate to pursue other interests.

Various investigators have tried experiments involving an electronic hookup, a kind of two-way underwater telephone system for dolphins in separate tanks so that the two animals could communicate back and forth without seeing each other. Meanwhile, scientists eavesdropped with tape recorders on the kind and frequency of phonations exchanged between the two animals. Observers noted that, while one animal would sound off, the other seemed to wait until it finished before answering. Eventually, however, when all of these experiments were reviewed, it was determined that the dolphins were responding to each others' whistle signals all right and could also be conditioned to respond to certain sound cues the experimenters used. But it was generally conceded that at least in these experiments the dolphins were not transmitting abstract information to each other through this telephone hookup.

Signature whistles may also convey the individual animal's emotional condition, for as the Caldwells found, animals under stress such as might be expected during periods of captivity, or during intense medical examinations by humans, or even during groundings, may break their signal or their phonations may include a "quaver," which has the same connotation of distress it would under similar stress circumstances in humans.

Efforts to study communication between dolphins are further complicated by their lack of the kind of external expressions commonly found in such land animals as chimpanzees or even dogs. A wagging canine tail, for example, speaks volumes. Even facial expressions are a form of "paralanguage"—the lips drawn back in a snarl, teeth bared, the low rumbling growl—and they leave no question in our minds as to the message the animal wishes to convey: "Watch out. I may attack you!"

But with the comparatively stiff-bodied, fixed expression so common to the dolphins, scientists involved in communication studies are limited when they look to the dolphin for any real "body language." And yet, these animals do possess a body language, subtle as it may seem.

The angle at which a dolphin may hold its body is meaningful to another dolphin. It may say such things as, "Watch out. I've had enough of your foolishness!" When one dolphin confronts another from the front, opening its mouth and arching its back, this, say observers, is a threatening gesture. Conversely, when a dolphin closes its mouth and turns its body sideways, this is interpreted as a sign of submission.

It is also believed that dolphins are capable of resolving differences in the pattern of their pigmentation, enabling them to identify species at a distance and individuals at close range.

Touching is another form of communication among the cetaceans; it is used by these aquatic mammals in much the same way it is by land animals. This is particularly true in the relationship and tactile bond beween infant and parent, as well as in the mating ritual.

It is unfortunate that almost all emphasis of our cetacean research has been directed largely toward trying to crack the mystery of their echolocation (sonar) system, the area of greatest interest to the Navy and therefore the area of greatest funding for research. Only in very limited and isolated cases has some experimental work been done in the area of communication, with studies designed to reveal the intelligence of these animals. Despite the many popular written and filmed treatments of this subject, we still have few hardcore facts to go on. But as we gradually begin to understand more about the extent of communication between these marine animals, the next question some scientists will ask is, "Can we communicate with them?"

9

Dolphins at Large

Pliny the Younger (A.D. 62–113) wrote, "I have met with a story, which, although authenticated by undoubted evidence, looks very like fable. . . . There is in Africa a town called Hippo, situated not far from the sea coast; it stands upon a navigable lake, communicating with an estuary in the form of a river, which alternately flows into the lake or into the ocean, according to the ebb and flow of the tide. . . . " The Roman author then went on to say that people of all ages enjoyed fishing, sailing, and swimming there, and one day, while swimming toward the opposite shore, a youngster met a dolphin which played around him and eventually allowed the boy to ride upon his back.

"The fame of this remarkable accident spread through the town, and crowds of people flocked around the boy. . . . The next day the shore was thronged with spectators, all attentively watching the ocean and the lake. . . . The dolphin appeared again and came to the boy, who together with his companions, swam away with the utmost precipitation. The dolphin as though to invite and call them back, leaped and dived up and down, in a series of circular movements."

This he continued to do for several days until the people got brave enough to enter the water and began playing with him themselves.

"They ventured, therefore, to advance near, playing with him and calling him to them, while he in return suffered himself to be touched and stroked. . . . The boy, in particular, who first made the experiment, swam by the side of him, and leaping upon his back, was carried backward and forward in that manner, and thought the dolphin knew him and was fond of him, while he too had grown fond of the dolphin. There seemed, now, indeed, to be no fear on either

193

side, the confidence of the boy and the tameness of the other mutually increasing; the rest of the boys in the meantime surrounding and encouraging their companion . . ."

Was Pliny's story fact or fable? Would wild dolphins, on their own accord, really make friends and play with humans? From a few rare incidences of a remarkably similar nature, we now know that the Roman was most likely writing the truth. Here's one example: In September 1965, as hurricane Betsy sliced a swath across the Gulf of Mexico to sideswipe Florida, Mississippi, and Louisiana, two bottlenose dolphins seeking refuge from the storm swam through narrow Philip's Inlet into large and partially brackish Powell Lake in northwest Florida. Several days later, the badly slashed body of one of the dolphins washed ashore. Those who found it suspected it was a victim of the storm.

As usual that winter, the inlet sanded in and the lake was landlocked. Local fishermen knew that the surviving dolphin was trapped in the lake, but everyone figured it would leave in the spring when the rain-swollen lake waters pushed out the sand and opened the inlet again.

Spring came, the inlet opened, but the dolphin did not leave the lake. People began seeing it more frequently swimming near shore, as if, some said, it was searching for someone. Soon, however, the dolphin stopped searching. It had found its first friend—a dog belonging to a local fish camp.

The dog stood on its owner's dock and barked lustily at the dolphin rolling playfully in front of the dock. This continued until one day people noticed the dog and dolphin splashing around together in the shallows near shore. The spectacle stopped grownups in their tracks. But not the kids. There was no holding them back. At first the fishing-camp children joined the dolphin and the dog in their frolicking. But soon, every youngster in the neighborhood was in on the action. Adults gathered on shore to watch and said it was quite a show—the dog, the dolphin, and all the kids whooping it up in the water together. No one ever remembered seeing anything like it before. By summer, a kind of love affair had begun between the wild but friendly dolphin and the people of Philip's Inlet. It was not a young animal, but a nearly full-grown *Tursiops truncatus*, bottlenose dolphin. And there was no doubt

This wild but tame dolphin so loved whirring propellers that it often got into trouble nudging boats to encourage occupants to start their outboard motors. He soon acquired the name "Nudgy."

that the creature was enjoying immensely all the attention it was getting.

That summer I heard about the dolphin from my old fishing friend George Brown, who lived near the inlet. George had named it "Nudgy" because it liked to nudge boats until they started their motors. At an outboard's first sputter and roar, the dolphin would whip around behind the boat, and with its bottlenose inches from the spinning propeller, would follow it everywhere. Whatever the attraction was, the dolphin loved it. When the boat stopped, Nudgy tried to push it into action again.

At first, Nudgy's infatuation with motorboats amused most people. Others, it infuriated. Those it bothered most were anglers with their boats anchored quietly over a good fishing spot. This bumptious dolphin knocked them off, not once, but repeatedly. To the dolphin it was a delightful game.

While most of the local people understood the dolphin's playfulness and good naturedly tolerated this bizarre behavior, many out-of-towners complained about it loudly to the fishing-camp owners. It seems that some of the fishermen's wives panicked, thinking they were under attack from a shark, which made matters worse.

After a while, when there was no letup of tourist fishermen voicing the same complaint, some of the fishing-camp owners began making complaints of their own. The dolphin was hurting business, they said. It had to go.

"Nonsense!" exclaimed almost everyone else. If anything, the dolphin was good for business. It attracted people, they argued. People came from all over to see it because it was a natural clown, one that learned tricks by itself that were just as clever as any Flipper did on television. Best of all, Nudgy was their very own! That's how most people felt around Philip's Inlet. And that's the way it went that year.

The next spring, before the tourists came, it was a different story. Some of the fishing interests on the lake got together and decided that something *had* to be done about the dolphin. He had to be removed. The overflowing lake had finally blown its plug at the mouth of the inlet. The molasses-brown brackish waters gushed full force back into the Gulf of Mexico. What better time than the present, they reasoned, to entice the dolphin back the way he had come? After all, what self-respecting creature of the sea really

Roughhousing it with a companion, Nudgy closes his eyes as he splashes back into the lake. He retaliated by swatting the boy in the seat with his tail.

wanted to live in the lake's half-fresh, tannin-stained waters when it had the whole crystal-clear Gulf of Mexico at its disposal?

Early the next morning, before the children came to play, a lone boatman pulled away from a fishing camp and gunned his powerful outboard motor raucously. Almost magically, the dolphin's sleek gray form broke the surface as it came bounding across the black mirrored waters of the lake.

The boat roared off. Perhaps anticipating the thrill of the chase, Nudgy raced after it.

Down the lake and under the highway bridge they went, boat and dolphin, the one barely keeping ahead of the other. Nor did they slacken speed at the end of the lake where it narrowed to cut a brown swatch through the beach of sugar-white sand that had blocked the inlet all winter long. Now, however, the channel did not meander as it would in the summer. Now it was a straight, deep, raging torrent that rushed out to blight the emerald sea with great brown splotches for miles along the coast.

Through this opening shot boatman and dolphin in tandem, the two driven even faster by the fury of the water as it formed short but violent waves at the mouth of the channel where ocean met lake. And there, in that long, uneven turbulence, the boatman lost sight of the dolphin. Still, he continued seaward, pausing only momentarily to gun his motor to make sure the dolphin heard and followed it in the stained water.

Finally, some distance from land, the boatman cut his motor to half speed and turned back, feeling sure the animal was free. Through the pass, across the end of the lake, under the bridge, and back to the fishing camp he went. Just as he tied up, he heard a commotion near shore. Glancing over, he saw two boys and a dog frolicking in the shallows with a disturbingly familiar large gray form.

In the next few weeks this scene played again and again. Same dolphin, different boatmen. The results were always the same. Each time, the dolphin dutifully followed the boat down the lake, through the pass and out to sea. Then it impishly whirled around and beat the boat back to the dock.

The boatmen were furious. The kids loved it. Almost everyone else heartily disapproved of the efforts to remove the dolphin. After all, they argued, he was their business, too. Moreover, he had made it perfectly clear that he preferred the company of people to those of his own kind. "Leave him alone!" they insisted. And they meant it.

But there seemed no way for peaceful coexistence between the fishermen and the dolphin. As the furor grew, so did Nudgy's problems. More than once the dolphin felt the wrath of irate fishermen when his friendly bump against their boat brought him a most unfriendly bump of their oar against his head. Soon, Nudgy's once smooth brow began to show the scrapes and scars of these encounters. But no one was really sure which were battle scars and which the result of his fascination with whirling propellers. Only once was there absolutely no doubt what had happened. Someone speared Nudgy.

It was a shocking, unthinkable cruelty. Fortunately, the wound was superficial and soon healed. But from then on, Nudgy wore the unmistakable marks of a five-pronged fish spear on his flank. Such malice was more than anyone could understand. The only one who seemed to bear no bitterness over it was Nudgy himself.

Soon, newspapers carried accounts of the wild dolphin that had befriended the people of Philip's Inlet. As Nudgy's popularity grew, more people came to see him. Even people from the adjacent states of Georgia and Alabama made special trips to this small Florida fishing community to pay their respects to the friendly dolphin. As his fame spread, it eventually attracted the attention of a commercial oceanarium that kept and trained dolphins for public display. These businessmen felt that Nudgy would be a natural attraction for their dolphin show. They came to Philip's Inlet to see if this talented creature could be caught and persuaded to go professional, joining the ranks of other dolphins that daily entertained crowds of people in exchange for free room and board.

Of course, there was no way to tell Nudgy that if he went professional he would never have to hunt for food as long as he lived. But the hard-working fishing folk of Philip's Inlet could surely understand the practicality of this arrangement.

Indeed they did. The people of Philip's Inlet understood perfectly. For the first time—some perhaps more reluctantly than others—both the prodolphin and the antidolphin people agreed. The dolphin was free to go wherever it wanted, but no one, they vowed, would ever take Nudgy from them.

Once or twice it seemed that some litigation might arise over the issue of who owned the wild dolphin and who had the right to claim it. But in the end the dolphin lovers of Philip's Inlet settled the issue through sheer determination.

One sunny afternoon I drove to the coast to catch Nudgy's performance. As usual, it was an informal affair taking place in the shallow water near shore where the youngsters cavorted with the big gray dolphin while the adults stood round in the shade of the loblolly pines watching the fun. Since I wanted to take pictures of the activities, I asked the kids to play close to a dock where I could look down on the action and photograph it.

Nudgy was on his best behavior. On cue from a suddenly started outboard motor, he shot across the intervening water to hover almost pantingly behind the idling motor, as if daring it to go. When it was slipped into gear, the boat leaped away with the dolphin seemingly glued by the end of its nose to the spinning propeller. Around and around they went in tight circles until the motorboat sputtered to a stop. With an adroit lift of his head, Nudgy tipped the foot of the motor out of water and the crowd applauded.

Closer to shore, a small boy summoned the dolphin for more fun and games by sticking the brass bell of an old taxi horn in the water and squeezing the rubber bulb to make it squawk.

Nudgy streaked back to the kids, flipped over on his back and swam circles around them upside down. The shore crowd cheered. As if taking a bow, Nudgy righted himself, swam between a youngster's legs, then surfaced to loll on top of the water in the middle of the group to accept good-naturedly the children's hands that patted him in praise of his fine performance. This kind of activity went on for the rest of the afternoon. Everyone seemed to be having a wonderful time. Especially the star performer.

Several months passed before I got down to the coast to see Nudgy again. But I kept track of his activities through short notes from my friend George. Once again trouble sprang up between Nudgy and the fishermen. Then abruptly, everything quieted down. I heard nothing from George.

Finally, after a while, I drove to the inlet to see what was happening. When I saw the dolphin, I knew things had changed. The fishing-camp owner whose dog and children had first befriended him now kept Nudgy in an enclosure bounded by docks. It was a kind of pen about thirty feet square. All that really confined the

Nudgy's apparent delight in a squawking horn sends him swimming in circles on his back.

dolphin in that shallow water was a chicken-wire fence strung around the dock pilings. The dolphin could easily leap over the low structure and escape whenever he wished. But apparently he chose not to do so.

I learned that the owner let him out at night to feed in the lake; he returned of his own accord. During the day he always stayed inside the enclosure. Surely, the fishermen had no complaints about this arrangement.

With other visitors, I walked out on the dock to see the dolphin. Instead of swimming around inside his pen as I had expected, Nudgy stayed away from the sides, remaining more in the middle of the enclosure, warily eyeing the people on the dock. Some children splashed water, trying to coax him closer so they could pet him. But the dolphin kept his distance, periodically lifting his head to breathe while watching the onlookers.

Only once did I see him respond to a spectator. An older boy tossed several pebbles toward him to attract his attention. Nudgy responded by lifting his head and squirting a sizable stream of water straight at his tormentor, scoring a direct hit. Everyone but the boy laughed. Then the crowd quickly dispersed. Most took it to mean that Nudgy was simply in no mood to play.

After the spectators left, I stood on the dock watching the dolphin. I felt depressed. It was not a happy scene. I had enjoyed seeing Nudgy free. Seeing him there by himself in the muddy water of his chicken-wire pen seemed wrong, even if he could come and go whenever he wished. Perhaps it was only my imagination, but at that moment I had the distinct impression that the dolphin's perpetual smile was no longer there.

Some scientists say dolphins are so well adapted to their environment that when coupled with an intelligence that may equal, or exceed our own, they may have achieved an existence far more perfect than man's. From the joyful expressions on the faces of these wild dolphins playing catch with jumping mullet, one could believe this to be true. Courtesy Marineland of Florida.

I never saw him again after that. Several weeks later, George sent me a short note. All it said was, "We lost Nudgy last night."

I never asked what happened. I do not want to know. I would rather believe that one night when he went out to feed, Nudgy just kept going down the lake, under the bridge, through the inlet, and out to sea.

Apparently, the only thing wild dolphins want from man is companionship. Dolphins in captivity perform for food rewards. But this is not the reason why Nudgy struck up relationships with man. Indeed, in instances such as the one described by Pliny the Younger in the beginning of this chapter, when dolphins were offered food from the hands of their admirers it was always refused. The dolphins chose to catch their own live fish from the sea. Evidently, food rewards were not the attraction. These marine mammals had nothing else to gain but the friendly, playful relationship itself.

While long-term relationships between wild dolphins and divers in the open ocean are rare, there are at least a dozen incidents a year

throughout the world that could be termed something more than mere chance meetings. It has happened in all the waters of the world, from the South Pacific to the Caribbean and the Atlantic. Off the coast of England for over three years a twelve-foot-long male bottlenose dolphin developed one of those rare close relationships with local divers. The British divers found this dolphin, whom they named Donald, caused such a stir with his friendly personality and playful antics that an organization called the International Dolphin Watch was established in the hope of documenting other such encounters throughout the world.

Man has always been fond of his domesticated animals. We make pets of the dog, the cat, even an occasional animal from the wild. We enjoy these animals on a certain level. A special bond develops between man and animal, whether the object of our affections is a venerable old tomcat, or in the case of Joy Adamson of *Born Free* fame, an African lion. For man to befriend, train, and relate to a domestic animal is one thing; to accomplish this with a wild animal is quite another. To pursue this feat into the alien world of the sea with one of its creatures takes on even greater significance. Moreover, when there is the possibility that this animal may possess an intelligence on a par with our own, then watch out!

The temptation to see ourselves and our human reactions mirrored in these relationships is overwhelming. How easy it is to anthropomorphize our smiling dolphin friends, this remarkable species of marine mammals that not only always appears to be happy, but extremely clever as well. So far, man has capitalized on this cleverness, this penchant for play, for his own amusement with dolphins in captivity. Perhaps now, by looking more closely at the animals under these conditions, we will see other interesting facets of their collective character. Perhaps we will begin to see behind the enigmatic dolphin smile to that indefinable "something" that has long nurtured the ever-growing dolphin mystique. Only then will we be able to understand why some scientists have selected the dolphin as their most logical choice for man's first breakthrough in intraspecies communication.

10

〰〰〰〰〰〰〰〰〰〰〰〰〰〰〰〰〰〰〰〰〰〰〰〰

Communicating with Dolphins

"To actually live with a dolphin twenty-four hours a day is a very taxing situation. Much more so than I had anticipated. Unlike a dog, unlike a cat, unlike a human, a dolphin is more like a shadow than a roommate. If given the opportunity, *he will never leave your physical being.*"

Thus wrote Margaret C. Howe in a special report detailing the events of living with Peter, a full-grown male *Tursiops truncatus* dolphin, in a special pool built for the two of them at the Dolphin Point Laboratory of the Communications Research Institute at St. Thomas, U.S. Virgin Islands, in 1965. This was one of the most unique experiments ever conducted to see if it were possible for humans to communicate with dolphins on a higher level than ever before achieved. This line of research originated with Dr. John C. Lilly, whose efforts from the mid-1950s on led to the development of the Dolphin Communication Laboratory at St. Thomas.

The initial proposal was for Miss Howe to live in close contact with Peter for a period of two and a half months. During this time, the human researcher was to attempt to establish a communication level with the dolphin that would include a personal relationship and she hoped, vocal exchanges.

By this time it was known that dolphins could mimic sounds they heard not only in their environment, but in ours as well. What impressed researchers, however, was that these mimicked sounds could be made out of water through the dolphin's blowhole and in frequencies which we humans could hear. It was believed by some that the reason the mammals produced these "humanoids," as Dr. Lilly came to call them, was solely for humans to hear. Certainly, they said, such sounds were not produced for their own kind. Thus began the research to determine just how adept a dolphin might be at developing this ability to mimic human speech. The Virgin Island laboratory seemed ideal for the experiment. Portions of the facility were modified in such a way and with great forethought to make both participants in this experiment as comfortable as possible. In the flooded "house" part, Miss Howe would sleep on a bed elevated inches above the water level where she could still maintain contact with the dolphin swimming in the two-foot-deep water of the main living area. This room opened by way of a half-Dutch-door arrangement into a flooded veranda containing adequate water for the dolphin to swim in. Strategically located microphones in the house unit would pick up all the vocal exchanges of the participants and would be tape-recorded. Miss Howe could also do her own cooking by a special kitchen arrangement in the area and could keep a written record of the events on a deck projecting out over the pool.

Before undertaking the two and a half-month experiment, Miss Howe spent seven days and seven nights living with Pam, a female dolphin, to see exactly what problems might need remedying before undertaking the long-term project. This particular dolphin had had several traumatic experiences that made her reluctant to engage in any close relationships with humans. According to Dr. Lilly, during the filming of the movie *Flipper*, a diver had speared the dolphin three times. For over two years now she showed all the symptoms of being a loner, apprehensive of humans and totally unresponsive to efforts made to encourage any kind of close relationship.

During the week of involvement with Miss Howe, however, the traumatized Pam gradually lost some of her fear of humans and developed a freer relationship. Thanks to Miss Howe's urging and skill, Pam reached the point where she could be fed by hand and was beginning to vocalize with the researcher, albeit in "Delphanese," dolphin chatter consisting of clicks, chirps, whistles, and other typically dolphin phonations, rather than the humanoids that mimicked our vocalizations.

Later, during the actual two-and-a-half-month-long experiment with Peter, there were times when Miss Howe wished fervently that

Dr. Lilly with three of his friends. Photo courtesy Marine World/Africa U.S.A. and Human/Dolphin Foundation.

she had engaged the more docile Pam in this long-term effort, rather than the rambunctious male dolphin that seemed mainly interested in continual play when he should settle down and learn his lessons with his young teacher.

That she was able to carry out this experiment in extreme isolation and at considerable mental and physical discomfort, during which the close relationship necessitated her being wet most of the time, is a credit to Miss Howe's dedication to the project.

At the beginning of the experiment, detailed in its entirety in Dr. Lilly's book *The Mind of the Dolphin* (Doubleday & Co., 1967), the dolphin Peter used only the Delphanese clicking at Miss Howe's request to learn such specific words as *ball*, *Bo-Bo clown*, or *toy*, while playing with the dolphin with these different items. Gradually, however, the dolphin began using a few humanoids, sounds that vaguely resembled words spoken to him by Miss Howe. For example, to the phrase "Hello, Peter," he might include in his clicking the humanoid "Oh" sound. Also, toward the end of the experiment it appeared that Peter could alternate his responses with the researcher's requests. When Miss Howe spoke, Peter seemed to listen, then would respond.

In conclusion, however, it appeared that to try to teach dolphins to mimic our speech was not the way to bring about a breakthrough. No one realized this more than Dr. Lilly himself, a man whom many consider the foremost authority on dolphins and proponent of the popular belief that dolphins are of equal or greater intelligence than man and that once we learn how, we will be able to communicate on the most abstract levels with them.

How did we come to this belief in the first place? The idea is not particularly new. It gained considerable momentum in the mid-1950s, when we learned that such animals as dolphins were capable of emitting sounds in the air that sometimes sounded strangely like the sounds they had heard in our world. We realized then that they were able to mimic sounds, imitating not only such things as a generator humming near their tank, but also human words that were spoken to them. Moreover, scientists such as Dr. John Lilly, while studying the anatomy of these marine mammals, were quick to notice that the dolphin's brain was astonishingly similar to the human brain and weighed only a bit more—1600 grams to our 1400-gram organ. No other animal on earth has a brain so similar to ours while maintaining a comparatively similar body weight. Was it possible, then, that this marine mammal with incredible sonar capabilities might be taught to share a give and take of thoughts and ideas with us? That was the first question that came to Dr. Lilly's mind.

All it seemed necessary to do was find a method, a means of breaking through, perhaps an Esperanto based on commonly understood symbols, some way to exchange information that would be comprehensible to dolphin and human alike. Only in the last twenty years have any serious efforts along these lines been attempted, and then only by a few scientists who believe strongly enough that dolphins possess the greatest potential for man and animals to talk together, to share thoughts and ideas in the same way as humans.

Fascination over the sounds dolphins make is not new. For example, consider this statement. "The dolphin, when taken out of water gives a squeak and moans in the air . . . for this creature has a voice and can therefore utter vocal or vowel sounds, for it is furnished with a lung and a windpipe; but its tongue is not loose, nor has it lips, so as to give utterance to an articulate sound (or a sound of vowel and consonant in combination). . ." This astonishingly astute observation was made not by one of today's scientists, but by Aristotle 2,380 years ago!

Today, the question is not so much *can* man communicate with cetaceans such as dolphins and killer whales, but how *well* can we communicate with them? We have communicated well enough in public oceanariums to be able to train the cetaceans to perform incredible feats on cue from humans. Now, one might well ask, is there more? Can we go beyond this point? What can we logically expect?

Dr. Lilly has asked himself these questions an untold number of times over the last twenty-five years. He comes from the finest academic background, but his methods and views have often been so radical that fellow scientists consider him something of a maverick, one whose ways and means of tackling the communication problem are a bit far off the well-beaten scientific path. A graduate of Cal-Tech and the University of Pennsylvania Medical School, by 1949 Dr. Lilly was deeply involved in human-brain research. Then, in 1955, he had his first contact with dolphins at Florida's Marine Studios, where a year earlier the then-director of the oceanarium, F. G. Wood, Jr., had recorded typical underwater sounds made by dolphins in captivity and published a description of them.

The whistles, creaking doorlike noises, barks, grunts, and rasping sounds of these mammals fascinated Dr. Lilly. But more fascinating than these was the realization that the dolphin's brain was so similar to the human brain. Soon, his research revealed some of the mysteries of how the dolphin produced these sounds and from which part of its head they originated. With dolphins that died in captivity, he relentlessly studied the anatomy of the various mysterious structures. And it was at this time that he had an opportunity to examine closely the dolphin's brain. This one organ—this mental computer—so impressed Dr. Lilly that it was to form the main thesis of his future arguments that dolphins are superior animals capable of providing a communication link between man and an alien intelligence. The dolphin brain was similar to the human brain in all of its structures, and it possessed an enormous amount of well-developed, convoluted cellular matter. What was an animal whose size so closely resembled that of man's doing with such a well-developed brain, wondered Dr. Lilly. How was it using it? Was it possible that dolphins were actually intellectually ahead of us and were communicating among themselves in such a sophisticated fashion that humans could not even comprehend how they are doing it? The possibilities were intriguing. Dr. Lilly's subsequent research revealed a great deal about the physical makeup of the dolphin and its extraordinary capabilities, hitherto unknown.

The 1950s and 1960s marked a period of strong interest in the underwater world. It was the beginning of the scuba era when man began taking his first look underwater with self-contained underwater breathing apparatus. It was a time when much was discussed and written about dolphins and a possible breakthrough in communication with them. While the scientific world thought along these lines, the public was caught up in the idea of the incredibly beautiful relationship that might develop between man and dolphin. And if there was any doubt just how great that could be, all one had to do was see it each week from one's own living rooms by tuning in on the extremely popular television series "Flipper." Who could help but believe it would all come true with this intelligent, happy-go-lucky, lovable Lassie of the sea?

It was a time for optimism, at least in this field. Twenty years ago Dr. Lilly optimistically predicted, "In the next decade or two, the human species will establish communication with another species: nonhuman, alien, possibly extraterrestrial, more probably marine; but definitely highly intelligent, perhaps even intellectual."

As the years passed, Dr. Lilly's popular books and papers kept both the public and the academic community apprised of his unorthodox research efforts to understand the dolphin and to try to establish a communication breakthrough. Prior to 1965, he wrote, "I felt that the scientific viewpoint of total objectivity, of the noninvolved scientific observer, was the be-all and the end-all for one's life. I am not convinced that such a dispassionate noninvolved view of ecology will ever work." It was bad enough that, to many staid scientists, his whole program sounded a bit too fantastic for words, especially scientific words. But some of it, they scoffed, bordered on science fantasy.

Dr. Lilly was later to write: "Any scientist who deviated from this traditional scientific procedure without consultation with his peers and colleagues in general was discredited by the scientific community. Any far-out new deductions, theories, or discoveries not subjected to the scrutiny of at least portions of the scientific community were considered to be not worth discussion, further investigation, or comment by the organized scientific concensus." He went on to add that if a scientist wrote a book that was not examined and published by an accepted scientific publisher but was instead written in everyday language and published for the general public, then that work was discounted.

In 1967, when his earlier efforts to develop a dolphin/human communication link proved entirely too slow and possibly not the best way to proceed, Dr. Lilly ventured into research areas where at least dolphins had never been before—he injected 100 micrograms of LSD-25 into a 400-pound dolphin and recorded its responses. These included a 10-30% increase in vocalization, especially when a human or another dolphin was in the tank with the subject. Summarizing these experiences, Dr. Lilly wrote:

> The important thing for us with the LSD in the dolphin is that what we see has no meaning in the verbal sphere. The meaning resides completely in the non-verbal contact exchanges. This is where our progress has been made in the last three or four years in developing this other level because we were forced to. We have had to do it in order to make any progress on the vocalizations and communications. In other words, we accept communication on any level where we can reach it. We are out of what you might call a rational exchange of complex ideas because we haven't developed communication in that particular way as yet. We hope to eventually.

This was written in 1967. In 1968, Dr. Lilly closed the dolphin research laboratory in the Virgin Islands and stopped all his research on these mammals. His explanation: "I did not want to continue to run a concentration camp for my friends, the dolphins." If it was not just his imagination, and dolphins were as he had found them to be, he said, then he felt there was an ethical problem in keeping them confined, a problem he was to discuss in more detail in a later book called *Center of the Cyclone*.

Primarily, Dr. Lilly wished to reevaluate all of his past experiences with dolphins and to write about his conclusions, not only with the dolphins, but in research upon himself through his own isolation in a watery environment. "You see," he said, "what I've found after twelve years of work with dolphins is that the limits are not in them, the limits are in us. So I had to go away and find out, who am I? What's this all about?"

Meanwhile, other researchers continued studying dolphins and learning more about how they echolocated and how man might benefit from this knowledge. But with Dr. Lilly out of the picture, research in the human/dolphin communication field was virtually nil. Hardcore, traditional scientists in an effort to keep all emotion out of their observations of these animals, found that they were caught in an odd predicament.

"No one can work closely with a dolphin without himself becoming emotionally involved," said one such scientist. It was an understandable human reaction. For who could daily look eye to eye with such an intelligent animal, one that constantly favored you with that perpetual enigmatic smile, and not feel an emotional bond with that animal?

Scientists wishing to carry on communications research where an emotional involvement might be a detrimental factor, were forced to avoid relating to the dolphins almost entirely. Others—the handlers, trainers, instructors—were given this chore of putting the test dolphins through their paces and having their emotional relationship with the animals, while the scientist directing the program, observed all progress at a distance by means of closed-circuit television.

Outside the academic community other equally diligent, if not quite so scientific, researchers experimented with human/dolphin relationships based almost entirely on an intimate, personal, emotional encounter with the animals. One such individual was Malcolm Brenner of Sarasota, Florida. Brenner's curiosity to learn about dolphins brought him to them. He began a close personal relationship with a female dolphin named Ruby. As far as Brenner was concerned, he achieved contact, achieved an intraspecies romance with this affectionate animal.

In an early episode involving his relationship with this dolphin, Brenner detailed an experience not uncommon among other dolphin observers who have had extremely close relationships with these animals as trainers, owners, or individuals seeking to go beyond the normal bounds of a trainer/dolphin encounter. Dr. Lilly first recognized the phenomenon in his work with dolphins in 1955, 1957, and 1958. In 1962 he described it as "a feeling of weirdness." As he said, it was the feeling that one was in the presence of something or someone of considerable intelligence waiting just on the other side of a nebulous barrier one was trying to penetrate. On that other side was an intelligent being that was trying just as hard as you were to breach the barrier, to communicate, and to reach you.

This is the essence of Brenner's observations, made quite accidentally and unscientifically. It occurred one cold March day when the dolphin enticed Brenner into her pool for an informal romp together. The water was ice cold to Brenner, who was in no mood to stay there very long. But Ruby seemed to be saying, "Come on in, the water's fine. We'll have a good time."

So in he went. There followed a tooth-chattering interlude for Brenner as the dolphin swam increasingly faster circles around him, almost frightening him with her violent activity. Ruby was apparently excited by his presence. She nuzzled him repeatedly, checked him up and down with her sonar, and whenever he reached out to stroke her, she allowed only so much familiarity, then retreated, just out of hand's reach, enticing him further into the pool. There, she nuzzled him, indicating that she was in the mood for play. But it was cold. Brenner stayed in the water as long as he could, then he had to get out.

Spotting her ball near the pool, Brenner decided it was a good opportunity to use Ruby's fondness for a game of catch as a reward to see if he could persuade her to mimic her name. He threw the ball to her and she quickly threw it back. They did this several times. The dolphin was enjoying the interchange enormously. Then, Brenner withheld the ball and said, "Come on Ruby, say Rooo-beee, like that. Come on. You can do it. Say Rooo-beee."

Instead, Ruby squeaked Delphanese in reply to Brenner.

"No, no," Brenner said. "That's not it. Say Rooo-beee." He held up the ball, withholding the reward.

As he continued urging her to pronounce her name, he suddenly became aware that her squawks had changed considerably. The sound was now distinctly two syllables. Nothing significant except that the two syllables were sounding strangely like those he had just been repeating for her. In response and as a reward for this, Brenner threw the cherished ball to the dolphin. In minutes he realized that she was beginning to copy the same speech pattern that he was using, even to the inflections of his voice, to mimic the very word he had been saying—"Rooo-beee."

Brenner was amazed that she had picked it up so quickly. Each time her pronunciation seemed less like his, he withheld the ball. Whenever Ruby replied with a sound that more closely imitated her name, he responded with the ball reward. He said, "We stood a few feet apart in the water of her pen, staring at each other intently with bright eyes and the excitement between us was palpable. Never in my life had I known such an intimate feeling of being in contact with an incredibly nonhuman creature. It felt like it was what I had been created to do. Our minds seemed to be running on the same wave. We were together."

What astonished Brenner most was that all of this had happened in less than ten minutes! After a while, however, it appeared that perhaps Ruby had become bored with the game. She pronounced her name as clearly as Brenner thought she could a couple of times, then stopped and babbled at him in Delphanese, accompanying this with a vigorous nodding up and down of her head, a gesture Brenner knew was associated with pleasure and which he called "ya-ya-ing." Then, she suddenly swam back a few feet, rose up out of water and emitted a peculiar noise that sounded to Brenner like "Keee-orr-oop" three times in short staccato deliveries.

Brenner said he did not know why, but it occurred to him to repeat the sound to Ruby, which he did as best he could. This effort was not as good as Ruby's, but she seemed to be expecting it of him. So Brenner tried. Ruby repeated the sound, but this time it too sounded different. She had modified it slightly. Brenner did likewise, responding by repeating the word with its modifications to

Ruby, again, as best he could with the inadequacy of human lips and vocal cords. She repeated the sound and again it was still a little different. Once more Brenner mimicked this difference, saying, "Kee-orr-oop."

"Suddenly," he reported later, "the light in my head went on. The sound I had just successfully imitated was the one she had been giving me in the beginning in response to my first attempts to make her say, 'Ruby'!"

Brenner realized this the instant the word was coming out of his lips and he said a whole bunch of fuses seemed to go off inside his head with the realization. He did a doubletake, staring at Ruby who seemed, he said, to be watching him with great concentration. "When she saw the doubletake and knew I knew, *she flipped out* and went ya-ya-ing around the pool, throwing water into the air very excited and apparently happy that this two-legged cousin of hers was progressing so rapidly."

What was the meaning of this experience, Brenner asked himself. Quickly, he reviewed what had happened. He had given the dolphin an English word—her name, Ruby—to pronounce. In response to it Ruby had replied with a Delphanese word or phrase that Brenner had at first ignored. But then through the ensuing interchange, during which he allowed himself to become the pupil rather than the teacher, there was some kind of progress. Brenner felt that each knew that the other knew what had transpired and it was this awareness, this "weird feeling," of Dr. Lilly's that had so impressed Brenner.

Certainly he knew that Ruby was smart enough to recognize her human name. But what was the significance of the sounds she repeated in return, he wondered. Could this possibly have been Ruby's Delphanese name for him? Or was it just the opposite; was she repeating her own Delphanese name? Brenner was fully aware that these were nothing more than his own projections of what he had heard. Still, it bothered him.

Months later, when he tried to coax Ruby to mimic her name for other observers, he was less successful. But she did manage to enunciate "Rooo-beee" once or twice. She seemed restless, impatient, as if being asked to do this mundane trick simply bored her. Brenner thought he understood why. He knew that dolphin trainers were aware of the dolphin's low threshold for boredom, that once they learned a trick they would repeat it only so often—then they were ready to go on to something new.

Years later, when Brenner told a scientist about this experience with Ruby, his contention was that it was "too bad he had not tape-recorded the interchange, that so often one hears what he wants to hear from an animal from which he is trying to elicit a favorable response. Besides," said the scientist, "one hears of this sort of thing so often . . ."

If this were the case, Brenner wondered why scientists did not try squawking Delphanese to dolphins more often. Maybe, in fact, humans were the real ones slowing down the communication between man and dolphin. Maybe it was not so much a case of their not being able to understand our language as it was our inability, and as yet, unwillingness, to try to understand theirs.

The idea that dolphins might indeed be able to teach us something was not just some observer's unscientific pipe dream. It was also the belief of Gregory Bateson, a seventy-year-old anthropologist, psychologist, and biologist who had worked with Dr. Lilly in the St. Thomas Dolphin Communication Laboratory during his early years there. Discussing his observations on a cetacean community, Bateson said, "A dog or cat takes a filial position with you, puts you in the position of parent or leader or whatever. . . . You don't get that when you are in the water with a porpoise, because you are the child and the porpoise is the parent. Now, if you are the swimmer and you can accept being the child and let the porpoise teach you, there's a lot you can learn about being a porpoise. Usually people think they learn about an animal by raising it, by becoming its parent; but then you are in a false situation, you are doing the leading. If you let the animal do the leading, the teaching, you will find out a lot more about the animal."

In an anecdote that pointed up the dolphin's reaction to boredom, Bateson told of an experiment set up by a psychologist involving a reverse learning study with a dolphin in Florida. The psychologist was trying to find out whether or not the dolphin could learn to learn. In this experiment the dolphin was trained to differentiate between two simple objects. One object meant to do this, the other object meant to do that. If the animal did the right thing, it was rewarded. If wrong, it was not rewarded. As soon as this test dolphin learned the difference between right and wrong, the psychologist switched the signals so that now the animal had to unlearn what it had just learned in order to be rewarded. It was not enough to do it right the first time. The psychologist set as a criterion fifteen right decisions on the part of the dolphin before he could come to a conclusion. Unfortunately, said Bateson, they could never get the dolphin to perform the same task that many times. The complaint from the psychologist's assistant was that, after performing the tests a number of times, the dolphin would then do it wrong and make a funny noise. Bateson asked if it had been recorded. No, replied the assistant, because the psychologist had not asked for that. To this day, said Bateson, we don't know the Delphanese for "Go to hell!"

After many years of observing cetaceans, naturalist and animal behaviorist Peter Warshall came to the conclusion that both man and whale seemed to have similarly large capacities for exchanging information based on its sequence using pitch and frequency variations. Humans, for example, accomplish this when using Morse code. Warshall believes, too, that the whale family may also be using a complex analog-communication system—that is, emitting sounds similar to the meaning of the data being communicated; for example, anger is communicated to others by the emission of a harsh, explosive, angry sound such as a cetacean might make by the sudden snapping shut of its jaws. Warshall believes that the mammals use digital speech when referring to "manipulable objects," while analog communication deals with more subjective things such as emotions. In this sense, says Warshall, "It may be that cetaceans have the ability [using their clicking apparatus] to communicate digitally as well as analogously."

While this kind of theorizing stimulates interest in breaking down the language barrier by letting an analog computer help humans solve some of the problems that have been baffling us until now, others are approaching the same problem from yet another direction. Their line of research is usually based on a method developed by Dr. David Premack, who since 1966, in a laboratory at the University of California, has been teaching Sara, a young chimpanzee, to communicate in the English language through the manipulation of variously shaped, colored pieces of plastic, each representing a word. Since Sara has been taught the meaning of each of these symbols, when they are placed on a magnetic blackboard, she correctly interprets and performs the appropriate actions requested of her. In response, she may place a certain sequence of symbols on the board requesting her trainer to perform some task. For example, Sara might sign "Mary-give-apple-Sara" and expect Mary, the trainer, to carry out the request.

Dr. Premack has progressed to the point where Sara understands more than just the meaning of a word symbolized by a plastic symbol. The complexities of her performance indicate that the chimp now understands basic sentence structure and is capable of handling relatively complex concepts. One of the most rewarding spinoffs of this line of research is that Dr. Premack's method is now being used with great success in communication therapy being done with severely retarded humans, who for the first time in their lives are able to communicate with people.

At Florida's Marineland, Bill Langbauer, a Boston University graduate student in marine biology, is using Dr. Premack's method to communicate with his test dolphins through the use of symbols. Langbauer's students are a pair of eleven-year-old dolphins—Betty and Sufi—captured about nine years ago. Sufi is a veteran in this line of research; she once worked for the Navy during seven years of sound-discrimination and behavior research carried on by the Caldwells.

In three thirty-minute sessions a day, Langbauer began teaching his dolphins the meaning of symbols he has devised representing different objects, persons, and actions. The symbols are white against black backgrounds on seven-inch-square metal plates; they have handles so that they can be easily held out and shown to the test animals by the trainer. For example, after a short session of patting and nuzzling his wards at poolside, Langbauer shows them one of the metal disks on which is printed a capital letter "V," instructing the dolphins to vocalize underwater. At the sound of Langbauer's starting bell, Betty does as she is instructed, responding with a rusty-door hinge kind of squeak.

Langbauer praises her and rewards his student with a handful of fish. Next, he shows Betty a plate painted with a white vertical line, the symbol to splash, which she does. Then, on to more advanced symbols, each eliciting the correct behavior. A white "O" requests her to turn her white belly up. Two pairs of white dots instructs her to stand on her tail. She may execute a dozen or more such exercises faultlessly, then apparently becomes bored and is reluctant to respond until coaxed by the trainer. Each correct response is followed by a food reward.

"It's hard to compare Premack's chimps with the porpoises because we've started with actions while he started with objects, but I'd say it took about the same time to teach a porpoise an action as it took to teach the chimp the object," said Langbauer, who obviously has no apprehension about the possibility of developing an emotional feeling for his students that might effect his tests. When asked if he feels personally drawn to Betty and Sufi, he answered with a definite yes, adding that porpoises are like people, some nice and some not so nice. Some may be playful and nuzzle your hand gently, while some may give it a hard whack.

Langbauer's interest in this line of research stems from a boyhood fondness for a pet dog and his wondering how the animal saw his environment in Langbauer's home. Today, Langbauer is equally curious about dolphins, wondering for example, whether their orientation enables them to know such things as left from right, north from south. Since the animals' sensitive sonar can differentiate between an aluminum and copper plate of the same color and size, Langbauer wonders just how much these animals know about each other or any human who may be in the water with them. Can they tell, for example, when another creature is sick or pregnant or simply annoyed?

"It's speculative, of course," said Langbauer, "but maybe he can tell us what his environment is like. The ocean is so vast and we are so limited in our mobility. The porpoise, though, can dive a thousand feet; he can see in the dark. He could be a diver's helper and the more he can communicate, the more help he would be."

At the Flipper Sea School at Florida's Grassy Key, trainers are using a similar method to communicate with their dolphins. Two white inverted "V" symbols on a black background in the center of a display board stand for "same as." Beside the symbol is a patterned seashell. The dolphin is then shown two seashells with different patterns, only one of which is similar to the example mounted on the board. The animal is asked to pick the shell that is the same as the one on the board.

Meanwhile, on the other side of the world at the Marine Mammal Research Laboratory of the University of Hawaii in Honolulu, Louis M. Herman, assisted by Douglas Richards, James Wolz, and other researchers, is involved in several dolphin studies now being undertaken in the Hawaiian Islands in an effort to learn more about not only how dolphins communicate, but how they relate to each other. Again, using variations of Dr. Premack's methods, these researchers have taught two dolphins to respond to as many as twenty-five "word" symbols in this artificial language, which the scientists feel will one day enable dolphins and humans to "talk" to each other.

One of the test animals already uses humanoid sounds above water to name objects in this artificial language. The scientists believe they have shown for the first time experimentally that dolphins are capable of understanding abstract word meanings and responding correctly when these words are used in new combinations.

At this laboratory the two dolphins are being taught the same thing in two different training modes—one employing computer-generated whistles, squeaks, and warbles that individually stand for objects and actions; and another in which these same objects and actions are expressed by trainer arm signals. In their watery environment the dolphin's sight is not nearly as acute as its hearing. Consequently, that portion of its brain that is believed to be used for processing what it hears is enormous compared to that part used in seeing. Trainers have found, however, that though the test animals are slow learners with visual cues expressed as displayed symbols such as Langbauer is using, they respond far more rapidly to arm signals. One would suspect, however, that if these animals were taught the same lessons in sonic signals they would prove the faster learners of the two groups.

Both test animals are female bottlenose dolphins captured two years ago in the Gulf of Mexico. Since individual dolphins have signal whistles which we believe are in effect their "names," the scientists tried to mimic these signature whistles when they named the animals. Therefore, the one learning arm gestures is named Akeakamai (Hawaiian for Lover of Wisdom) and the other, being trained in the acoustic language, is named Phoenix.

Lessons include certain objects familiar to the dolphins—a plastic hoop, a piece of plastic pipe, a ball, a Frisbee. During the training program these objects are floated on the surface of the pool, and either through a sonic or an arm signal, the dolphin is instructed to pick out a certain object, which he does by touching it with his nose.

If he has chosen correctly, an acoustic yes signal sounds and he is rewarded with a morsel of fish. If he chooses incorrectly, he gets a no signal and the trainer turns his back to the dolphin, a mild rebuff that is taken quite seriously. If one or the other of the animals should make two consecutive mistakes, either animal is liable to lift its head out of the water and squawk angrily. They may even throw a temper tantrum, picking up any one of their toys and hurling it violently and with uncanny accuracy at their instructor. This performance is quite

Anything in the dolphin's immediate environment becomes fair game. This one is playing with a sea turtle at Marineland of Florida. Courtesy Marineland of Florida.

indicative of the animal's low threshold for boredom and its all-too-human response. Normally, however, lessons progress without many of these interruptions and the animals receive considerable reinforcement and praise for their progress. The trainers accomplish this through hand-clapping and head-patting accolades.

After such lessons, in which the dolphin learns the sound or arm signals for various objects, these are then used in simple four-word sentences in which the dolphin is asked to perform some simple activity. For example, a combination of signals such as "Phoenix-Ball-Fetch-Gate" tells that animal to carry the ball to the gate that separates the pool. Such combinations as "Fetch-Right-Pipe" or "Fetch-Left-Pipe" have taught the dolphin the difference between right and left.

The signal "Hoop-Through" instructs it to swim through a hoop. In one instance when there was a swimmer in the water with the dolphin and on a whim the instructor signaled the unfamiliar command "Person-Through," the dolphin promptly responded by pushing the swimmer through the gate.

Elaborate steps have been taken at the laboratory to avoid any possibility that extraneous cues might tip off the dolphins as to how they are to respond during the performance of a lesson. Therefore, a researcher sits in a tower near the pool where he can randomly select on a keyboard the sonic signals given to the dolphin. Similarly, from this tower a researcher tells a poolside trainer which objects to throw into the pool. The trainer then blindfolds himself with opaque goggles so that his eyes will not inadvertently tip off the dolphin to which target he is going to be expected to touch. Nor does the trainer himself know until the instant the tower researcher instructs him to give the appropriate arm signal directing the dolphin to one of the objects.

Mark Sharp, a graduate student, has been training Akeakamai in the visual arm signals. If Sharp wants to tell the dolphin to fetch the ball to him, he will raise his arms over his head and drop them for the visual signal "Ball." "Fetch" is an extended arm sliced into the chest and "Person" is two arms extended in front and scissored up and down.

While Phoenix, the other dolphin, has been trained to perform the same task through the use of sonic signals, it is interesting to note that, when one sees an oscilloscope pattern of these acoustic words, "Phoenix-Ball-Fetch-Person," the pattern on the instrument's screen is similar to the arm signals used by the trainer.

Dr. Herman, the laboratory's director, said that one of the goals of these experiments was to try to better understand the roots of human language, to learn if it could have evolved from more primitive nonverbal processes and if this capacity for such communication is unique to man. He also hopes to learn how well these animals, which possess sonic capabilities far superior to man, can learn a language and how great a potential exists in this area.

According to Kenneth Norris, who directs the University of California studies involving dolphins, there has to be a reason why the sound-processing portion of a dolphin's brain is so highly developed. It seems designed to play a roll far more important than the

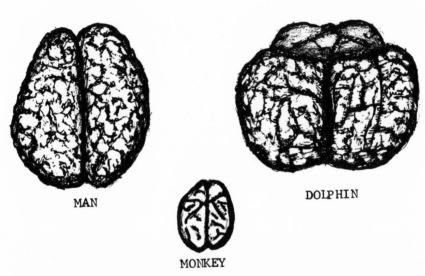

MAN   MONKEY   DOLPHIN

Comparison of brains drawn from actual photographs show strong similarities between human and dolphin brains. Some scientists believe that the well-developed, highly convoluted dolphin brain indicates a high degree of intelligence. One suspects they were not given that much gray matter just to carry on a relationship with sea cucumbers. Drawn from photographs by Dr. John C. Lilly, *Lilly on Dolphins*, Anchor Press/Doubleday, New York, 1975.

mere echo-ranging capabilities of which we know it is capable. But this explains only part of it, for bats too depend largely on this kind of sonar and yet this portion of a bat's brain is relatively small.

Scientists like Norris, and others in this field, want to know why dolphins possess such a large brain. "In an ocean full of dullards, what good is such a brain?" asks Norris. "Certainly, complicated nervous machinery is not needed for concourse with jellyfish, sea cucumbers, and sponges. Some of its capacity clearly relates to catching prey (including coordinated herding and attack), but there is clearly more than that involved."

In an effort to observe and relate to the social behavior of wild dolphins, one of Norris's California students, Jody Solow, swam daily with a school of wild dolphins that came into the bay of one of the Hawaiian Islands. At first she tried to relate to a school of ten dolphins off the nearby island of Lanai. Hoping to attract their attention, she performed underwater somersaults and other acrobatics. But the dolphins showed only mild interest in her activities. As the school turned to leave, Miss Solow groaned disappointedly in her snorkel. The sound came out sounding something like a "dog yawning," she said. To her surprise, one of the retreating dolphins immediately turned back and swam up beside her.

From then on Miss Solow used the sound each day to announce her presence to the school of dolphins; and the leader brought the entire school around her until they almost touched her. Miss Solow hoped that some day she would be as accepted by these dolphins, which now number some two hundred, just as Jane Goodall was accepted by a group of chimpanzees during her behavioral studies in Africa.

With varying degrees of success, others have tried to interest wild dolphins in their presence. The most recent attempt was documented on a public-television program entitled "Dolphin." It described how Steve Gagne and others used an electronic pipe organ invented by Gagne to try to attract the schools of spinner dolphins in the waters of the Bahama Islands. This method harks back to the early acoustic experiments of underwater pioneer Hans Hass in the Red Sea. From these and other similar efforts, we know that something about the rhythmic sounds of music played underwater seems to have a calming and attracting influence on marine creatures, especially cetaceans. During the Gagne experiment, in which he used scuba equipment and an air-powered keyboard organ on which he played plaintive tunes, audiences were delighted to see the almost choreographed responses from the dolphins swimming in unison with Gagne in a kind of slow-motion sinuous underwater ballet. If music hath charms to soothe the savage beast, then this was surely a fine example of it.

Psychiatrist and ecologist Dr. Sterling Bunnell, who in addition to his medical practice teaches evolutionary ecology at the California College of Arts and Crafts, tells us that our ability to gather a lot of information with our eyes compares to the dolphin's ability to swiftly gather a lot of information from his sense of hearing. "Dolphin language apparently consists of extremely complex sounds which are perceived as a unit," explains Dr. Bunnell. "A whole paragraph's worth of information might be conveyed in one elaborate instantaneous hieroglyph. For them to follow our patterns of speech might be almost as difficult as for us to study the individual picture frames in a movie being run at ordinary speed. It is not surprising that captive dolphins at first seem more interested in music than in the human voice. Our music is more similar to their voices than our speech is." And this may well be the reason why dolphins are attracted to our musical sounds.

The first not only to consider the possibility of developing some kind of electronic "black box" that would enable humans to manipulate sound transmissions between humans and dolphins, but actually to build such a box, is reputed to be the late D. Wayne Batteau of Tufts College. Though primarily a mathematician, Batteau's interests revolved around electronics and sound. Eventually, it was not mathematics but this interest in electronics and sound that got him involved with dolphins. He built the black box he dreamed of and it was ingenious, an electronic marvel that took the human voice and turned it into different frequencies of whistles—the same kind of whistles that dolphins exchange underwater. And the sounds from Batteau's box could be broadcast underwater in the same way. The remarkable box broke the human voice down so that the vowel sounds spoken normally in the back of the mouth, such as "o" and "u," were reproduced as low-pitched whistles. Those normally produced in the front of the mouth, such as "a," "e," and "i," were changed into high-pitched whistles. And since Batteau's box worked better with vowels than consonants, he generally chose Hawaiian words because they contained so many vowels.

In the experimental program with dolphins in which Batteau used his electronic box, it became a kind of underwater Premack method based on auditory whistle signals instead of plastic symbols. The researcher would speak into a microphone hooked up to the box and the resulting whistles would be broadcast underwater, creating an audible sound resembling the whistle language of the dolphins. The researcher might say "Kali," or whatever the name of the test dolphin happened to be, and that animal, hearing the sonic signal for its name, stood by attentively waiting for the next command.

The next words would instruct the dolphin as to what it was supposed to do, such as "Fetch-Ball." The dolphin would not budge until given the final word to do so—an electronically whistled "Okay." Immediately, the dolphin performed the task. If the instructor requested "Kali-Repeat-Ball," the dolphin responded by mimicking the whistle signal it had learned for "Ball," then touched the object, if it had not already done so.

From this point on the experiments became more complicated. Messages usually always contained the words "Do" and "Repeat," but to these were added perhaps a dozen more action words that the dolphins learned and were apparently becoming quite proficient in understanding and using for the task they were being asked to perform.

This line of research employed, for the first time, human speech converted to sounds which were surely more familiar to dolphins. For the first time, too, the mammals were being asked to respond in this more familiar mode underwater rather than in air, where they had once been asked to mimic what surely must have been far more difficult pronunciations for them, and to do this with the so-called humanoid sounds through their blowholes.

The possibility of using a form of whistle language, one that through some miracle of electronic wizardry could be understood almost instantly by two participants who lacked a common means of communication (the man and the dolphin), started others thinking about what marvelous things might be accomplished in making a possible breakthrough in a kind of artificial Delphanese language. Was it possible that we had been asking these presumably highly intelligent animals to actually lower their communication capabilities to our far more simplistic form of painstakingly stringing words together like beads until a whole string of such vocal symbols were necessary to create a single basic sentence, a single statement of fact?

What if it were possible that dolphins were so much more advanced than we that in their language they were able to communicate large chunks of information, literally volumes, in the space of a fractional second—the time it took to emit a certain kind of ultra-high-frequency sound wave? If that were the case, no wonder they were finding it so difficult to gear themselves down to our comparatively primitive level of communication.

Imagine the reverse of this situation—you are the alien expected by the dolphins to try and mimic their language in their environment. This would mean that, in order to obtain a food reward, you would have to go from your environment into theirs, stick your head underwater, and listen attentively while a trainer dolphin emitted a complex sequence of whistle signals so difficult for you to mimic that you would be lucky if you could whistle even one note that sounded like the original.

On the other hand, if that dolphin, through the magic of electronics, was able to convert his whistle language into electronic tones resembling English words and these were broadcast to you in your own air environment, how much easier for you at least to begin to try to understand what those words meant.

The next logical progression along this line of thinking is, of course, not Batteau's primitive black box, but one of our more modern miracles—the computer. Brains are themselves the ultimate computers. The thought of using a mechanical brain, a computer, to help establish a communication link with dolphins strongly appealed to brain specialist Dr. John C. Lilly.

In 1976, eight years after he had ceased all dolphin research to study himself, Dr. Lilly returned to the field of dolphin research. In the intervening years, Dr. Lilly's work and his popular writings about dolphins had finally born fruit in the form of such organizations as Greenpeace, Save the Whales, the Dolphin Embassy, and other similar groups that were actively involved in educating the public about the need to protect these forms of life.

While he had been away, practically no research based on his early 1955 to 1968 efforts had been done in the communication field, either of dolphins alone or between dolphins and humans. After he had terminated the research program in 1968, Dr. Lilly had not ceased thinking about the dolphins and the problems of interspecies communication. The same year he came out of "retirement," he formed the Malibu, California–based Human/Dolphin Foundation, a nonprofit organization formed to tackle the problem of communications with dolphins by using an entirely new approach. It was one you might have expected from a man like Dr. Lilly.

His methodology now would include some of the most sophisticated equipment in the business. He would use minicomputer hardware coupled with software programming techniques. The project was to be named JANUS, for Joint Analog Numerical Understanding System. Like the two-faced god of Roman mythology, this electronic JANUS was to face in two opposite directions, one toward humans, the other toward dolphins. It would be the go-between, the master mechanical brain intended to translate English words into sonic signals that would fall in the optimum auditory range of the dolphins. It was Batteau's black box amplified to its maximum capability, plus a few other innovations such as video screens that allowed both man and dolphin to see the sounds that were being transmitted through an underwater loudspeaker.

Long before JANUS was an actuality, Dr. Lilly had hypothesized the whole idea. He believed that, through the use of modern computer technology, it was possible to create a machine that would instantly change human-voice frequency into the dolphin's frequency range and in return transform their high-frequency responses down into the lower ranges more audible to humans. He spoke of it as being a doorway that must be opened between the human sonic box and the dolphin sonic box before a feasible route would exist that would take advantage of special "vocoders" that would function on each side of JANUS.

These vocoders were simply a technical design problem that could be solved, he thought, providing sufficient funds could be found, along with the special technical expertise it would take to do the job. Described simply, each vocoder would employ analog methods to analyze sounds it heard, then either multiply them on the human side up to ten times to put them into the dolphin's

Dr. Lilly and his wife, Toni, operating the JANUS computer that may provide the link enabling man and dolphin to communicate with each other on the most sophisticated levels. Photo courtesy Marine World/Africa U.S.A. and the Human/Dolphin Foundation.

frequency band or, in the case of a dolphin's response, divide by a factor up to ten, reducing the frequency down to the human level. And it would do this without any delay, which would mean there would be a free interchange of request and response so that each side would learn rapidly. In 1976, when this was still simply an idea, Dr. Lilly estimated that the design and construction of this equipment would cost one hundred thousand dollars.

An alternate method he envisioned involved the use of high-speed minicomputers and microprocessors that included equipment and programming capable of performing much as the vocoder; however, this particular combination would be operated with a keyboard, which would facilitate certain frequency modifications. The big advantage of this method was the flexibility of programming. The 1976 cost for this model was estimated to be about three hundred thousand dollars.

Ultimately, the funds were found and, by December 1979, the equipment was completed in a mobile laboratory that Dr. Lilly and his assistants could take to wherever the test dolphins were available. The system consists of a computer video terminal on the side being used by the human operator, with an underwater portion consisting of hydrophone, loudspeaker, television monitor, and video camera on the dolphin's side. On the terminal keyboard the human operator types out any number of letters from A to Z or numbers from 0 to 9. When the operator strikes a return key the computer broadcasts to the dolphin through the underwater loudspeaker the sound it creates from each of the typed characters. This sound is picked up by JANUS's underwater microphone, sent back through the computer, and simultaneously it creates a large picture character of the sound on television screens seen by both the dolphin and the human operator.

When the dolphin responds, his sounds are picked up by the computer's underwater microphone, and once again, JANUS transforms the sound into a visual picture on the two television screens. In this way the dolphin can see his own sounds by watching the underwater television screen, which helps him learn the elements of the sounds. Since the dolphin can see himself on the television screen, it also helps him to position himself properly before the instrument's hydrophone and loudspeaker.

To be sure of an adequate record of all that transpires, the machine provides a computer printout, along with a videotape recorder with two video and two sound channels in addition to an audio tape recorder. The audio and visual recorder misses nothing. It records all the screen characters, all the images of the dolphin underwater, and all sounds emanating from JANUS and from the dolphin. The computer printout provides details of sound amplitudes, frequencies, analytical data, the JANUS parameters in use, and the characters being created by JANUS as initiated by the human or the dolphin operator.

The researchers began by running tests to determine how well the dolphins could learn new sonic codes containing distinctive computer-generated signals. The mammals were first tested to see how well they could whistle and recognize sound signals of different frequencies and duration. From these experiments, Dr. Lilly and his team selected forty-eight frequencies found to be those easiest for the dolphins to discern. Generally, these frequencies were about ten times higher than those used by humans when they talk together: 3,000 to 40,000 Hz (cycles per second) for dolphins, as compared to 300 to 4,000 Hz for humans.

As the work progressed, Dr. Lilly and his group were able to program the computer's frequency analyzer and wave-form generator in a variety of ways to test the dolphin's reaction to a large number of alternatives to their initial procedure. Therefore, by January 1980, Dr. Lilly learned that the dolphins were not only interested in the program, but were responding enthusiastically with their whistle signals and showing signs of modifying their normal delphanic sounds toward our chosen parameters. Their learning is expected to take many weeks of exposure to JANUS, said Dr. Lilly. "We are still in the initial month of working with the dolphins and of modifying our programs in consonance with results obtained."

The Human/Dolphin Foundation (P.O. Box 4172, Malibu, California 90265) is a nonprofit, scientific research and educational organization whose program has been supported both by monetary contributions from friends, members of the foundation, and a few private foundations and by the contributions of professional time and effort of dedicated programmers, engineers, scientists, artists, actors, musicians, film people, television professionals, secretaries, and others who believe strongly in the project. Dr. Lilly says that the current expense required to keep the program moving is six thousand dollars a month.

He anticipates the next step of the JANUS project, once the dolphins master the whistle code created by the human operator with the aid of the computer, to be for the animals to learn various combinations of sonic-code elements, which will become meaningful to them as they perform the various behavioral tasks requested by the chain of signals.

The dolphins will be trained to use this artificial whistle code for their own purposes—to activate various devices connected to the computer. For example, by sounding certain code-whistle combinations, the dolphins can activate an automatic fish dispenser, start tape recordings of music, or start video tapes that will be played back to them for short periods underwater. They will be able to request certain responses from the computer operator, says Dr.

Lilly, "by controlling the printer in which various things can be spelled out in human language by means of the computed transforms of the teaching code into printed human language. [They] can demand the presence of a human away from the computer in which the human is in physical proximity to the dolphin. [They] can use the computer to synthesize human speech utilizing modern human-speech synthesis programs and output devices that are available commercially."

At this writing, these achievements are hypothetical, but Dr. Lilly feels that such things are now within our grasp, and granted sufficient time, energy, money, and interest, they will be achieved and the goal will be well worthwhile.

"I feel very strongly that the reward to the human species of such a program will be very great—beyond anything that I or anyone else can imagine," says Dr. Lilly. "Alternatives to human language and communication with another species is a program that may be able to capture as much human interest around the planet as we currently devote to human warfare. . . . The ancient extraterrestrials are here, waiting for us to grow up and maturely communicate. Let us stop destroying them and us and start a new evolutionary interspecies dialogue."

If any one man is capable of achieving this in our time, that man is Dr. John C. Lilly.

Bibliography

Airapet'yants, E. Sh. and A. I. Konstantinov. *Echolocation in Animals*. Translated by N. Kaner. Jerusalem: Israel Program of Scientific Translations, 1973.

Albro, Frank D. "Breakthrough! Shark Repellent." *Oceans*, September 1975.

Alpers, Antony. *Dolphins: The Myth and the Mammal*. New York: Houghton Mifflin Co., 1961.

Bain, Robert. "Ichthyological ESP, Fact or Fancy?" Unpublished manuscript.

Baldridge, David H. *Shark Attack*. New York: Berkley Publishing Co., 1975.

Barton, Charles. "The Navy's Natural Divers." *Oceans*, Vol. 10, No. 4, July-August 1977.

Beebe, William. *Half Mile Down*. New York: Duell, Sloan and Pearce, Inc., 1934.

Burgess, Robert F. *The Sharks*. New York: Doubleday and Co., 1970.

——. "Shark vs Porpoise." *Science Digest*, June 1971.

Caldwell, David K. and Melba C. "The Dolphin Observed." *Natural History* 77(1968):58-65.

——. "Dolphins Communicate—But They Do Not Talk." *Naval Research Reviews*, Vol. 25, Nos. 6 and 7, June-July 1972.

——. *The World of the Bottlenosed Dolphin*. New York: J. P. Lippincott Co., 1972.

Clark, Eugenie. *The Lady and the Sharks*. New York: Harper and Row Publishers, 1969.

——. "Flashlight Fish of the Red Sea." *National Geographic*, Vol. 154, No. 5, November 1978.

"Conference on the Shark-Porpoise Relationship." Washington D.C.: The American Institute of Biological Sciences, 1967.

Cousteau, Jacques Yves. *The Ocean World of Jacques Cousteau, Invisible Messages*, Vol. 7. The Danbury Press, New York: Harry N. Abrams Inc., 1975.

Davis, Flora. *Eloquent Animals: A Study in Animal Communication*. New York: Coward, McCann and Geoghegan, Inc., 1978.

De Carli, Franco. *The World of Fish*. Translated by Jean Richardson. New York: Abbeville Press, 1978.

Devine, Eleanore and Martha Clark. *The Dolphin Smile: Twenty-Nine Centuries of Dolphin Lore*. New York: The Macmillan Publishing Co., 1967.

Earle, Sylvia A. "Humpbacks: The Gentle Whales." *National Geographic*, Vol. 155, No. 1, January 1979.

233

Eibl-Eibesfeldt, Irenäus. *Land of a Thousand Atolls: A Study of Marine Life in the Maldive and Nicobar Islands*. Translated by Gwynne Vevers. Cleveland and New York: The World Publishing Co., 1966.

Faulkner, Douglas. "The Ocean Community: Lesson for Man." *Oceans*, Vol. 4, No. 2, March-April 1971.

Fish, Marie Poland and William H. Mowbray. *Sounds of Western North Atlantic Fishes: A Reference File of Biological Underwater Sounds*. Baltimore and London: The Johns Hopkins Press, 1970.

Frey, Hank. "The Not So Silent World." *Sea Frontiers*, Vol. 9, No. 1, February 1963.

Griffin, Walker I. "Making Friends With a Killer Whale." *National Geographic*, Vol. 129, No. 3, March 1966.

Hartman, Daniel S. "Florida's Manatees, Mermaids in Peril." *National Geographic*, Vol. 136, No. 3, September 1969.

———. *Behavior and Ecology of the Florida Manatee, Trichechus Manatus Latirostris (Harlan), at Crystal River, Citrus County*. Ph.D. dissertation, Cornell University, 1971. University Microfilms International, Ann Arbor, Michigan.

Harvey, E. N. *Bioluminescence*. New York: Academic Press Inc., 1952.

Hass, Hans. *Challenging the Deep*. Translated by Ewald Osers. New York: William Morrow and Co., 1973.

Hendrickson, Jr., Walter B. "Voices of the Deep." *Sea Frontiers*, July-August 1977.

Hodgson, Edward S. and Robert F. Mathewson, eds. *Sensory Biology of Sharks, Skates and Rays*. Arlington, VA: Office of Naval Research, Department of the Navy, 1978.

Hoyt, Erich. "Orcinus Orca." *Oceans*, Vol. 10, No. 4, July-August 1977.

Hyman, Ann. "Whales: Why Do They Sing?" *The Florida Times-Union* Jacksonville Journal, May 4, 1980.

Idyll, C. P. *Abyss: The Deep Sea and the Creatures That Live in It*. New York: Thomas Y. Crowell Co., 1964.

Kasanof, David. "Living Sonar." *Sea Frontiers*, Vol. 8, No. 3, 1962.

Lilly, M.D., John Cunningham. *Lilly On Dolphins: Humans of the Sea*. New York: Anchor Press/Doubleday, 1975.

———. *Communication Between Man and Dolphin: The Possibilities of Talking with Other Species*. New York: Crown Publishers, Inc., 1978.

Linehan, Edward J. "The Trouble With Dolphins." *National Geographic*, Vol. 155, No. 4, April 1979.

Lubow, Arthur. "Riot in Fish Tank 11." *New Times*, October 14, 1977.

Marx, Robert F. *Secrets Beneath the Sea*. New York: Belmont Tower Books, 1958.

McBride, A. F. and D. O. Hebb. "Behavior of the Captive Bottlenose Dolphin." *The Journal of Comparative and Physiological Psychology*, Vol. 41, No. 2, April 1948.

McIntyre, Joan, et al. *Mind in the Waters: A Book to Celebrate the Consciousness of Whales and Dolphins*. New York: Charles Scribner's Sons, 1974.

McMahon, Jim. "Schooling Phenomena Observed." Personal communication to the author. February 1980.

McNally, Robert. "Echolocation." *Oceans*, Vol. 10, No. 4, July-August 1977.

Murphy, Geri. "Discover the Dolphin." *Skin Diver*, Vol. 28, No. 8, August 1979.

Norris, Kenneth S., Editor, et al. *Whales, Dolphins and Porpoises*. Berkeley and Los Angeles, California: University of California Press, 1966.

Ommanney, F. D. and the editors of *Life*. *The Fishes*. New York: Life Nature Library, Time Inc., 1963.

Payne, Roger. "Swimming With Right Whales." *National Geographic*. Vol. 142, No. 4, October 1972.

———. "Humpbacks: Their Mysterious Songs." *National Geographic*. Vol. 155, No. 1, January 1979.

———. "At Home With Right Whales." *National Geographic*. Vol. 149, No. 3, March 1976.

Perry, Richard. *The Unknown Ocean*. New York: Taplinger Publishing Co., 1972.

Perry, Robert. "And Man Created A Great Fish: Killer Whale." *Science and Mechanics*. Winter 1979.

Pliny the Younger. *Letters*. Translated by William Melmoth. New York: P. F. Collier and Son, 1909.

Reiger, George. "Dolphin Sacred, Porpoise Profane." *Audubon*. January 1975.

Ridgway, Sam H., ed., et al. *Mammals of the Sea: Biology and Medicine*. Springfield, IL: Charles C. Thomas Publisher, 1972.

Riedman, Sarah R. and Elton T. Gustafson. *Home is the Sea: For Whales*. New York: Rand McNally and Co. 1966.

Roessler, Carl. "Color Control, Multihued Fishes." *Oceans*. No. 5, September 1977.

Scheffer, Victor B. *The Year of the Whale*. New York: Charles Scribner's Sons, 1969.

———. "Exploring the Lives of Whales." *National Geographic*, Vol. 150, No. 6, December 1976.

Scott, Robert Falcon. *Scott's Last Expedition*. Boston: Beacon Press, 1957.

Sea Secrets. Question regarding Delaware Bay fish sounds, Vol. 4, No. 5, September-October 1970.

Sebeok, Thomas A., ed., et al. *How Animals Communicate*. Bloomington and London: Indiana University Press, 1977.

Shark Research Present Status and Future Direction. ONR Report ACR-208, Arlington VA: Office of Naval Research, Dept. of Navy, April 1975.

Tarpy, Cliff. "Killer Whale Attack." *National Geographic*, Vol. 155, No. 4, April 1979.

Tinbergen, Niko and the editors of *Life*. *Animal Behavior*. Life Nature Library. New York: Time, Inc. 1965.

Walker, Theodore J. "The California Gray Whale Comes Back." *National Geographic*, Vol. 139, No. 3, March 1971.

"Whale Talk: Song and Dialect." *Science News*, Vol. 117, No. 2, January 12, 1980.

Wood, Forrest G. *Marine Mammals and Man: The Navy's Porpoises and Sea Lions*. New York: Robert B. Luce, Inc. 1973.

Young, William E. with Horace S. Mazet. *Shark! Shark! The Thirty-Year Odyssey of a Pioneer Shark Hunter*. New York: Gothan House, 1934.

Index

237